S0-CRV-032

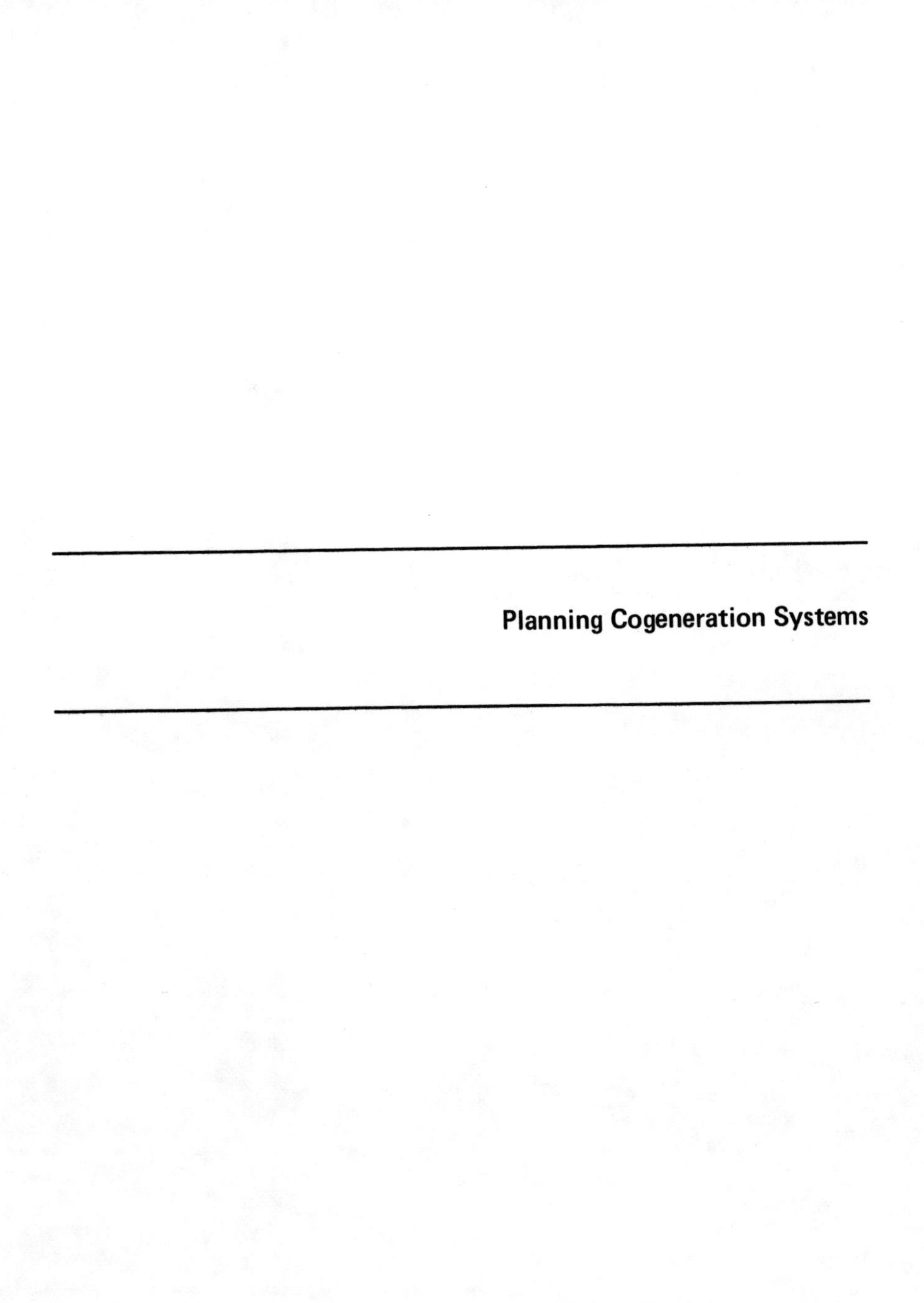

# Planning Cogeneration Systems

# Planning Cogeneration Systems

***Dilip R. Limaye***

THE FAIRMONT PRESS, INC., P.O. Box 14227, Atlanta, GA 30324

*Planning Cogeneration Systems*
by Dilip R. Limaye



Printed in the United States of America

Library of Congress Catalog Card Number: 84-48107
ISBN: 0-915586-95-9

Library of Congress Cataloging in Publication Data
Main entry under title:
Planning cogeneration systems.

Includes index.
1. Cogeneration of electric power and heat.
I. Limaye, Dilip R.
TK1041.P53 1984 621.1'9 84-48107
ISBN 0-915586-95-9

# Contributors

A number of experts in the field of cogeneration have contributed to this book. The efforts of the following individuals are gratefully acknowledged.

Thomas K. Casten, President, Cogeneration Development Corporation
James R. Clements, President, United Enertec, Inc.
Craig R. Cummings, Asst. Vice President, Science Applications, Inc.
Harry Davitian, President, ENTEK Research, Inc.
Thomas R. Germani, Vice President, Johnson and Higgins
E. Wayne Hanson and Douglas D. Ober, CH2M Hill, Inc.
Fred H. Kindl, President, Encotech, Inc.
John M. Kovacik, Manager, Cogeneration Systems Sales, General Electric Company
F. Richard Kurzynske, Program Manager, Gas Research Institute
M. P. Polsky and R. J. Hollmeier, Fluor Engineers, Inc.
Beno Sternlicht, Chairman, Mechanical Technology, Inc.
Robert G. Uhler, Vice President, National Economic Research Associates
M. V. Wohlschlegel, G. Myers, and A. Marcellino, Westinghouse Electric Corporation
James J. Zimmerman, President, Cogeneration Society of New York

# Foreword

Cogeneration represents a classic case of how changing economic conditions can give an old technology new life. It has been practiced since before the turn of the century, but had declined steadily in importance for decades. The dramatic events of the 1970s placed energy efficiency in a new light and have led to a great resurgence of interest in the last few years. Cogeneration will continue to increase in importance in the coming years because it makes good economic sense and it helps to meet important national energy conservation goals.

An early example of cogeneration was at the Dow Chemical Company's Midland, Michigan plant. Herbert H. Dow founded the company in 1897 to extract bromine and chlorine from brine using a new electrolysis process. Initially he generated his own power with wood-fueled steam engines without recovering the waste heat from the engines because there was little need for steam in the original plant. Then in 1910 he developed a new chlorine process which required the concentration of brine in massive vacuum evaporators.

At the time Dow's powerhouse engines exhausted steam at 150°F after generating power. Since the new process required great quantities of heat, Dow routed the wasted steam to the brine evaporators, allowing him to save fuel and to avoid the cost of installing a separate boiler. Cogeneration, as this practice has come to be known, has been utilized continuously by Dow Chemical since that time. Today the company is probably the world's largest cogenerator.

Around the turn of the century most industrial electricity users, like Dow Chemical, generated their own power. At the time electric utility service was unreliable, expensive, and not widely available, so on-site generation was usually a better alternative. In the early 1900s over half of all the electricity used in the United States was self-generated, much of this from cogeneration systems.

However, the relative importance of power generation by industry declined steadily through the late 1970s. The primary reason was the improvement in service by electric utilities. As electric generation technology advanced, larger and more efficient power plants were built, which lowered the cost of power. As the electric utility industry grew, more and more industrial plants were served by utilities. In the 1920s and 1930s electricity generation became a regulated business, and on-site generators could be regulated as a utility if they sold any excess power. These many factors combined to cause cogeneration to decline greatly in significance, reaching a nadir of only 3% of the total electrical power generated in the U.S. around 1980.

In the last few years, though, there has been an enormous resurgence of interest in cogeneration. The most important factor in this development has been the huge escalation of energy prices over the last decade. The oil shocks of 1973 and 1979 caused fuel prices to skyrocket. Likewise, electric rates escalated greatly due to technological limits on the maximum size of power plants, environmental regulations, nuclear problems, and higher fuel prices.

Cogeneration has always been able to save fuel, but as long as energy prices were cheap, spending capital to save energy could not be economically justified. With today's high prices, cogeneration can represent one of the best investments that an energy user can make. Future price trends, as electric rates continue to rise and finite resources of fossil fuels are depleted, will make cogeneration increasingly attractive.

The passage of the Public Utilities Regulatory Policy Act in 1978 and the favorable decision by the Supreme Court in 1983 also played a critical role in cogeneration's renewal. PURPA removed most of the institutional obstacles to cogeneration, which had previously caused many potential cogenerators to disregard it despite excellent economics. Cogenerators can now count on fair treatment by the local electric utility with regard to interconnection, back-up power supplies, and the sale of excess power.

In addition, cogenerators no longer fear being regulated as an electric utility. With PURPA it is now possible for energy users to install the most efficient cogeneration systems, especially those with high electric outputs, allowing the full potential of cogeneration to be realized on a national scale.

There have also been a number of other significant incentives to cogeneration. Improved technologies have been developed, which are more efficient or are lower in capital cost. The gas turbine combined cycle cogeneration system, first installed in the late 1960s, is now being widely implemented, whereas in the past most had been steam turbine systems.

Other parties have created incentives for cogeneration. The Federal and state governments have encouraged cogeneration to help meet energy conservation goals and to contain future electric rate increases. Many electric utilities want more cogeneration in their service territories in order to avoid spending scarce capital on expensive new power plants and to provide needed generating capacity.

Cogeneration is highly capital intensive, so its proliferation depends to some extent on the development of techniques for project financing. Cogeneration lends itself well to project financing: it generates large revenues or cost savings; substantial tax benefits are realized from accelerated depreciation and investment tax credit; the nature of the assets allows a high degree of debt financing; and there is little technological risk. Because of these characteristics, cogeneration has caught the fancy of the financial community in the last year or two, and many firms are actively seeking cogeneration projects to finance.

Yet another spur has been the development of a cogeneration industry, in which third-party project developers finance cogeneration systems and sell power and heat over the fence to large energy users. In effect, these firms serve as nonregulated utilities to the energy user, and, since they are taking advantage of the high efficiency of cogeneration, they can sell energy at a discount below the cost of alternative sources of supply.

This development will greatly increase the number of potential applications for cogeneration, making available the cost savings to energy users who lack capital or who do not want to be in the power business themselves.

The future of cogeneration is indeed very bright. The further development of industrial cogeneration capacity is well documented and will occur rapidly in the next few years. One study has estimated the market for cogeneration equipment and services at $60 billion by the year 2000. Another survey predicts a doubling of industrial cogeneration capacity by 1993.

Many new cogeneration technologies are in the developmental stage and will be commercialized in the 1980s and 1990s. Gas turbines with higher firing temperatures will increase the potential fuel savings and further improve the economics of these cogeneration systems. Fuel cells allow electricity to be generated without the theoretical limitations of thermal cycles, and the waste heat can be put to use to create highly efficient cogeneration systems on a very small scale.

Phosphoric acid fuel cells will be commercially available at competitive costs in the late 1980s, and molten carbonate and solid oxide fuel cells are likely to be available in the 1990s. Coal gasification, while not a cogeneration technology per se, will allow the most efficient cogeneration systems to burn coal-derived fuel rather than being limited to oil or natural gas, eliminating the most significant drawback of gas turbine and diesel cogeneration systems. These new technologies, combined with projected energy price trends, will further enhance the importance of cogeneration.

Most cogeneration today is practiced in large-scale, industrial plants. The recent development of low cost, prepackaged cogeneration systems in small sizes is opening up the market for relatively small energy users, with total energy bills as low as $250,000 per year. These prepackaged units can be produced cheaply since they will be manufactured in large quantities, and the economies of mass production will compensate for the loss of economies of scale resulting from the small size. They can be installed quickly, require little engineering, and are easy to service.

By reducing the size at which cogeneration becomes economic, these prepackaged systems will exponentially expand the number of potential sites for cogeneration. Prospective cogenerators will include anyone with a substantial demand for both power and heat, such as: hospitals, colleges, shopping centers, multi-family residential buildings, and small industry. It has been estimated that 10,000 MW or more of small cogeneration capacity will be installed in the next decade. In response to this demand, there will develop a small cogeneration industry, which will own and operate small cogeneration systems and will sell power and heat to these energy users.

The economic attractiveness of cogeneration is most dependent on the price of electricity, so it will first be widely implemented in regions with high rates, such as California and the Northeast. Other areas which are served by a utility with nuclear problems will also develop cogeneration capacity rapidly. As electric rates continue to escalate along with other energy prices, cogeneration will spread across the entire U.S. Eventually there will be few new central station power plants built, and most new electric generating capacity will consist of cogeneration systems.

The time has come when it is no longer economical for large users of electricity and heat to use power generated by conventional condensing plants. To survive in today's fast-moving world, all major energy users will eventually be forced to cogenerate to stay competitive.

Gerald L. Decker, *President*
*Decker Energy International, Inc.*
*Winter Park, Florida*
*March, 1984*

---

Gerald L. Decker was one of the principal investigators of the Energy Industrial Center Study, sponsored by the National Science Foundation. This study was the first significant analysis of the new role of cogeneration as the nation's energy future, following the Arab oil embargo.

# Contents

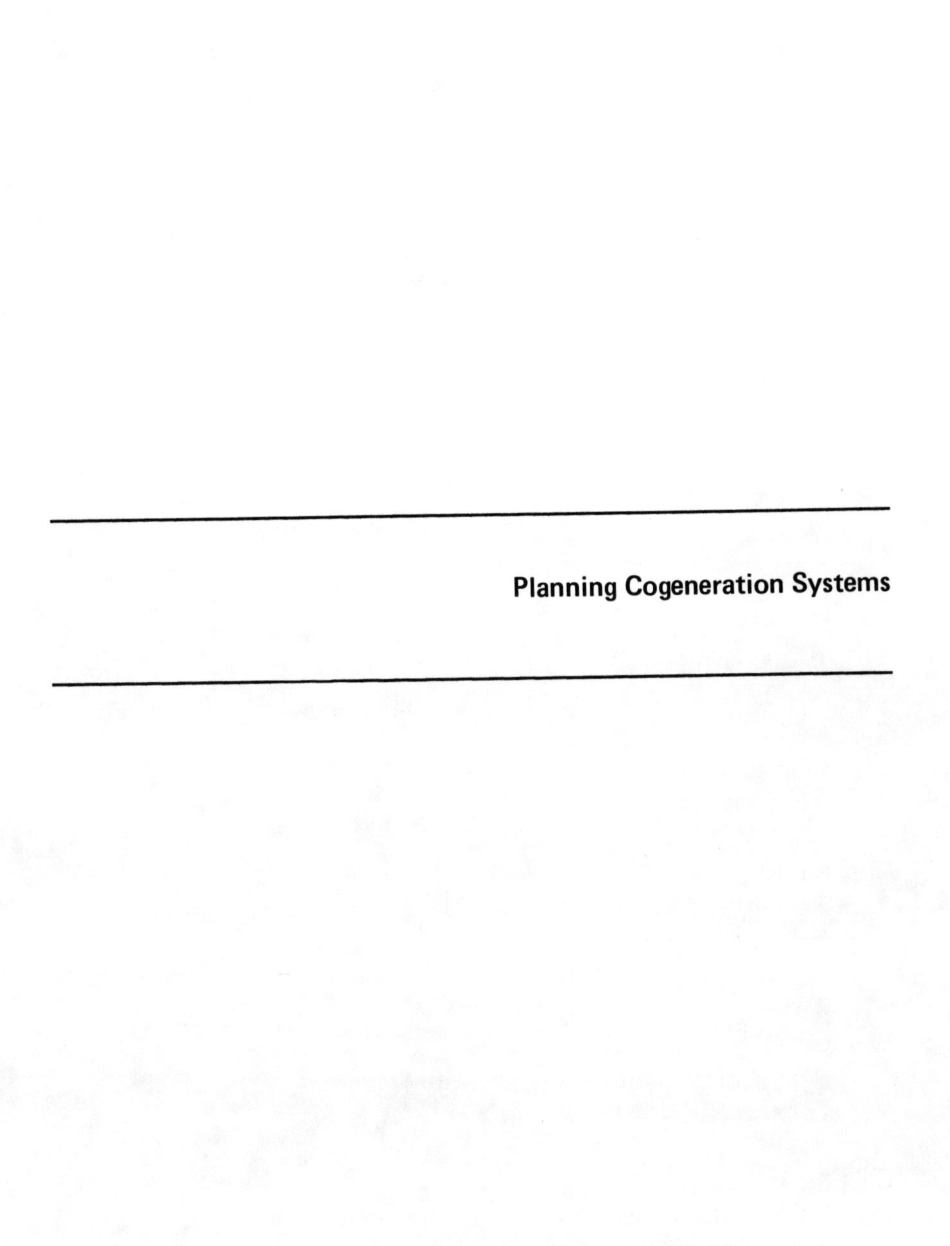

# Planning Cogeneration Systems

# Chapter 1

## Cogeneration in the 1980s

**Dilip R. Limaye**

According to the General Accounting Office, U.S. industry and electric utilities use nearly half the primary energy consumed, and the waste heat from power generation and process energy use amounts to over seven million barrels per day oil equivalent.[1] Cogeneration can offer a method to reduce the amount of waste heat by simultaneously producing electricity and useful thermal energy from a common primary energy source. Because of its potential for efficient use of energy, cogeneration is receiving increasing attention in the U.S.

The concept of cogeneration is not new. Industrial generation of electricity has been practiced for a long time. In the early 1900s, most industrial plants generated their own electricity and approximately half of this was using cogeneration.[2] On-site generation/cogeneration was more reliable and less expensive than utility generated power. However, in the 1920s and 1930s, the regulation of electric utilities, first by state agencies and then by the Federal government, resulted in elimination of unproductive competition, and consolidation and extension of utility service areas. Coupled with the availability of inexpensive fuels for power generation and technological advances in central station utility generation and transmission of electricity, industrial generation/cogeneration became economically less attractive. From the 1920s to the mid-1970s, there was a generally declining trend in the proportion of electricity cogenerated in industry.[3] Other factors contributing to this declining trend included the following:

- Industry was hesitant to invest in generation because of the possibility of Federal and state regulation as a utility, and the related reporting requirements.
- Utilities offered very low prices for excess power sold by an industry to a utility.
- Utilities charged high prices for standby or supplemental power needed by the cogenerator.

As a result, industrial generation declined from 18% of total electric generation in 1941 to about 4% in 1977.[4]

### The Changing Energy Situation

In the last decade, the energy situation in the United States has undergone a significant transition. The nation has faced increasing prices and decreasing availability of conventional energy sources, energy supply disruptions, environmental constraints to the utilization of coal, and high capital costs for expanding the energy delivery system. Efficient utilization of our energy resources has become a very high priority and cogeneration has become economically attractive. At the same time, Federal legislation has attempted to remove some of the institutional barriers to cogeneration and small scale power production. Moreover, the problems faced by electric utilities have resulted in increased interest on their part in industrial cogeneration.

The electric utility industry is beset with financial problems of unprecedented magnitude. New generating plants committed in the 1968–1974 time frame, when demands were forecast to grow at an annual rate of 7–10 percent, have been mostly deferred or cancelled. The basic problems faced by the utilities include:

- High costs of new capacity
- High interest rates
- Escalating fuel costs
- Environmental/siting constraints

- Increased customer resistance to rate increases
- Regulatory lag.

These problems, coupled with slower load growth, have led to lower revenues than forecast, while the capital requirements for new capacity have continued to escalate rapidly. Utilities have therefore been forced to borrow large amounts of money at very high interest rates. Utility earnings have been depressed, and it is becoming difficult to offer new stock (except at prices below book value) *and* to cover the interest costs of new borrowing.

**Power Generation in the 1980s**

Many utilities, looking ahead to the late 1980s, see their best prospects in (a) completing plants now almost completed, and (b) to some extent discouraging increases in load growth with the expectation that a two percent annual growth rate will be manageable, allowing time for their economic situation to stabilize before having to undertake another new plant. As part of this basic approach, all utilities would find it advantageous to flatten their system load curve, and to reduce or eliminate use of expensive peaking generation requiring use of high cost fuels in relatively inefficient power plants. Cogeneration could contribute significantly in this approach. In addition, utilities may be able to raise capital through innovative financing schemes such as joint ventures or third-party arrangements to build new capacity for cogeneration.

Until recently, few utilities actively encouraged or participated in cogeneration. In a recent survey of utilities, conducted by EPRI as a part of case studies of industrial cogeneration,(5) utilities expressed concerns regarding the potential loss of baseload, reliability and maintenance of cogeneration equipment, interconnection costs, the availability of cogenerated power when needed by the utility, and the need for standby capacity.

The significant changes in the economic and institutional aspects of power generation, which occurred in the 1970s and are expected to continue in the 1980s, have created a trend towards increased interest in and acceptance of industrial cogeneration by utilities. These changes have led utilities to consider industrial cogeneration in their planning for future capacity needs, and have also resulted in the growing recognition of cogeneration systems as a utility business opportunity. Cogeneration ventures, owned and operated by a utility, can be highly complementary to traditional utility operations, and possibly offer a potential for higher profits than the traditional utility business. Utilities are therefore increasingly interested in examining opportunities for participation in industrial cogeneration projects.[6]

It appears that the changing economic and institutional environment will lead electric utilities in the 1980s towards a gradual redefinition of their traditional role. In the future, utilities are likely to be seen as "energy service companies" rather than merely as suppliers of electricity. In this new role, utilities may embark upon a number of new types of business ventures, some of which have already been undertaken by utilities in the past several years. Thus, utilities may actively encourage cogeneration and may even participate in such projects.

## Cogeneration—Old Game, New Rules

A number of significant changes have occurred in the last few years relative to the institutional and regulatory aspects of cogeneration. The National Energy Act (NEA) of 1978 contains a number of important provisions which attempt to remove institutional barriers to cogeneration. The most important provisions are in the Public Utility Regulatory Policies Act (PURPA), which provides the following for facilities that "qualify" by meeting certain operating and efficiency requirements.[7]

- Utilities must purchase any and all power that the qualifying facility (QF) wants to sell.
- The rate offered by the utility for such power purchase should be based on the "avoided cost" of the utility.
- The rates charged by a utility to a QF for standby/backup power must be nondiscriminatory.
- The QF is exempted from utility regulation under the Federal Power Act, the Public Utility Holding Company Act and state regulations related to rates and financial reporting.

A qualifying facility must not be more than 50% owned by an electric utility.

Federal Court cases in Mississippi (which ruled PURPA unconstitutional) and in the D.C. Court of Appeals (which asked FERC to reconsider the 100% avoided cost rule and the requirement for utilities to interconnect with a QF) had created some uncertainties in PURPA implementation. However, both of these rulings were appealed to the U.S. Supreme Court, which upheld the PURPA legislation and the FERC implementation rules. After the Supreme Court decision, most states have completed the implementation of PURPA rules.[8]

In addition to PURPA, three other parts of the 1978 NEA also provide incentives for cogeneration. The Powerplant and Industrial Fuel Use Act (FUA) allows cogenerators to be exempted from prohibitions on the use of oil and natural gas. The Natural Gas Policy Act (NGPA) provides an exemption from incremental pricing of natural gas to cogenerators. The Energy Tax Act (ETA) provides a 10% investment tax credit for certain property which may be used with cogeneration systems. Also, additional incentives were provided in subsequent legislation passed by the 96th Congress.[9]

### Implementing Cogeneration Projects

The state regulatory environment relative to PURPA implementation has led to cooperative efforts among industry, utilities and third party investors for financing and implementing cogeneration projects.[10]

The reasons for considering such cooperative efforts are:

- Cogeneration is likely to be more capital intensive than a conventional energy system, and industry may have other uses for capital which are more attractive.
- Industry is hesitant to make major cogeneration investments because of perceived uncertainties relative to PURPA.
- Industry may not have the skilled staff needed to operate and maintain a power generation facility.
- Industry may not consider power generation a natural extension of its primary business, even when such generation is economically attractive.

- Utilities are generally willing to accept a lower rate of return than industry.
- Industrial plant managers may be hesitant to face the problems related to the handling, storage and use of coal (the preferred fuel for cogeneration) and the associated environmental requirements.
- Utilities can offer the necessary expertise in the construction, operation and maintenance of cogeneration systems.
- Third party investors are attracted by the tax credits and other financial benefits offered by cogeneration ventures.

Many utilities are currently actively seeking cooperative ventures with industry. Thus, industries interested in cogeneration may find the local utility a willing and cooperative partner.

**Options for Cooperative Efforts**

A number of options exist for cooperative efforts among industry, third parties and utilities to implement cogeneration, including:

- Sole utility ownership of the cogeneration plant with sale of thermal energy by the utility to industry.
- Joint venture between industry and utility (with utility owning 50% or less to qualify the cogeneration facility for PURPA benefits).
- Third party ownership with contracts for thermal energy and electricity sales to industry and utility respectively.
- Partial ownership with the utility owning the power generation equipment and industry owning the remaining plant.
- Sole industry ownership but operating control (dispatch) by utility.

The main theoretical justification for a multi-party approach is to share the risk of a project. This reduces the total risk to any one participant, while commensurately reducing the possible returns. In addition, a joint venture arrangement should reduce the "moral risk" of a project where two or more participants must cooperate: if all participants have a stake in the operation, they will all have an incentive to do their part. This is particularly appropriate in the case of cogeneration, where cooperation between the industrial user(s) of the thermal energy and the utility purchaser of the electricity is essential.

**Financing Cogeneration Projects**

Cogeneration projects can be financed through a variety of joint venture arrangements.

The term joint venture refers to financing a specific contractual relationship or undertaking among two or more participants in a project. The legal relationship among the participants may take the form of a partnership, a joint owned corporation, or an unincorporated association. The partnership structure offers great flexibility for joint venture arrangement between utilities and industrial cogenerators, because the partnership agreement can be quite flexible.[11]

In general joint venture financing structures are of two types: undivided interest and project entity.

The undivided interest structure may be used by a sole owner or a joint venture. The owner(s) of an undivided interest in a project contributes capital in proportion with his ownership and receives profits in the same proportion. Any funds borrowed to capitalize the project would be shown on the participant's balance sheet as would the assets of the project. If a wholly owned subsidiary owns an undivided interest in the project's assets, the parent would have to consolidate the subsidiary's accounting with its own as if it owned the undivided interest itself.

Under the project entity approach, the cogeneration project would be established as a separate entity to own and operate the equipment. Project financing is the term used to describe the raising of capital to finance a project entity approach. The cogeneration project entity may be owned by a sole owner or by a joint venture.

Each participant contributes capital and receives project benefits based on agreement with other participants (not necessarily in proportion to his ownership interest). Capital is secured by the assets and future cash flows of the project. The debts of the project are not shown as debts on the participants' balance sheets. The project entity approach may affect the participant's cost of credit or ability to raise additional debt even though the project debt does not appear on the participant's balance sheet.

The practical considerations involved in selecting a financing arrangement include: the ability of each participant to raise capital, the cost of the capital, alternative uses for the capital and the ability to maximize and utilize the tax benefits and other incentives available to the project. The participants must identify their financial capabilities and constraints to determine their proper roles in any cogeneration project. The structures identified above may be used to finance projects that could not be undertaken by any one participant or may produce better financial results by reducing capital costs through the best use of tax credits and other incentives.

## Prospects for Cogeneration

The changing economics of cogeneration have led to a great deal of activity and interest on the part of both industry and utilities to evaluate and implement cogeneration projects. A large number of applications have been filed with FERC for certification as a qualifying facility.[12] An analysis of these shows that the greatest activity is in California and the southwest.

A recent assessment of cogeneration potential for the U.S. Department of Energy estimated a total of over 42,000 MW.[13] Table 1-1 shows a summary by industry group and region. While these estimates may appear to be high (they were developed using a 7% ROI criterion), they nevertheless indicate the likelihood that substantial activity related to cogeneration is likely to be undertaken in the U.S., particularly if the uncertainties in PURPA are resolved in a manner favorable to cogeneration.

Because of the efficiency of cogeneration systems, they offer economic benefits to both the industry and the utility. Cogeneration results in conservation of our energy resources and lower environmental impacts. Cooperative efforts between industry and utilities can lead to the implementation of the optimum cogeneration systems, providing benefits to industry, utilities and society.

**Table 1-1. DOE Estimates of Industrial Cogeneration Potential**

A. BY INDUSTRY GROUP

| *SIC* | *(MW)* | *(%)* | *Number* | *(%)* |
|---|---|---|---|---|
| 20 | 7,146 | 17 | 863 | 27 |
| 26 | 8,414 | 20 | 454 | 14 |
| 28 | 9,800 | 23 | 408 | 13 |
| 29 | 10,976 | 26 | 179 | 6 |
| 33 | 2,823 | 6 | 307 | 10 |
| Remaining Sector | 3,665 | 8 | 920 | 30 |
| | 42,824 | | 3,131 | |

B. BY REGION

| *Region* | *Number of Potential Plants* | *Potential Power Generation* (MW) | *Potential Electricity Generation* ($10^6$ Kwh) | *Potential Steam Generation* ($10^6$ 1b/yr) | *Potential Energy Savings* ($10^9$ Btu/Yr) |
|---|---|---|---|---|---|
| New England | 289 | 3,014 | 17,464 | 98,843 | 115,386 |
| NY/NJ | 265 | 2,833 | 19,070 | 116,035 | 128,872 |
| Mid-Atlantic | 319 | 4,536 | 30,183 | 215,531 | 206,834 |
| South Atlantic | 544 | 5,757 | 40,464 | 396,778 | 294,648 |
| Mid West | 559 | 5,226 | 37,874 | 321,993 | 251,377 |
| South West | 335 | 11,362 | 91,714 | 763,314 | 631,891 |
| Central | 186 | 2,411 | 17,895 | 153,122 | 119,403 |
| North Central | 38 | 506 | 4,072 | 33,817 | 27,684 |
| West | 408 | 7,708 | 43,219 | 216,761 | 278,744 |
| North West | 150 | 1,316 | 8,642 | 64,474 | 58,830 |
| TOTALS | 3,093 | 44,669 | 310,593 | 2,380,634 | 2,113,620 |

Source: *Cogeneration World,* March/April 1982.

## Acknowledgment

This chapter is based upon the author's research as part of a project sponsored by the Electric Power Research Institute titled Evaluation of Dual Energy Use Systems (DEUS), Project RP1276.

## References


1. U.S. General Accounting Office, *Industrial Cogeneration—What it is, How it works, Its potential,* EMD-80-7, April 29, 1980.
2. Fred J. Sissine, "Energy Conservation: Prospects for Cogeneration Technology," Issue Brief No. IB81006, Library of Congress, Washington, D.C., February 1982.
3. Frederick H. Pickel, *Cogeneration in the U.S.: An Economic and Technical Analysis,* M.I.T. Energy Laboratory Report, MIT-EL-78-039, Boston, Massachusetts 1978.
4. Synergic Resources Corporation, *Cogeneration Case Studies and Data Base,* Report prepared for the Electric Power Research Institute, forthcoming.
5. Synergic Resources Corporation, *Industrial Cogeneration Case Studies,* EPRI EM-1531, Electric Power Research Institute, Palo Alto, CA, September 1980.
6. Thomas C. Hough and Dilip R. Limaye, *Utility Participation in DEUS Projects: Regulatory and Financial Aspects,* SRC Report 7070-R1, Bala Cynwyd, PA, 19004-1105.
7. For further details, see Final Rules Implementing Sections 201 and 210 of PURPA, 45 *Federal Register,* 12214 and 17949.
8. Dilip R. Limaye, "Overview of State Regulation of Cogeneration." Paper presented at the Workshop on Cogeneration: Federal/State Regulation, Arlington, VA, April 1983.
9. The legislation providing additional incentives to cogeneration includes the Crude Oil Windfall Profits Tax Act, the Energy Security Act and the Housing and Community Development Act of 1980.
10. Dilip R. Limaye, "Evaluation of Industry/Utility Cooperative Efforts in Cogeneration," Paper presented at the Fifth World Energy Engineering Congress, Atlanta, GA, September 1982.
11. John F. Barry, "Limited Partnerships and Municipal Financing of Cogeneration Projects," Paper presented at the Third International Conference on Cogeneration, Houston, TX, October 1983.
12. "FERC Applications for Qualifying Facility Status," *Cogeneration World,* March/April 1982.
13. "Industrial Cogeneration Potential (1980-2000)," *Cogeneration World,* March/April 1982.

# Chapter 2

## Overview of Planning and Prefeasibility Considerations

***James J. Zimmerman***

Cogeneration systems are planned for only one of two reasons: To save money or make money. Given that either savings or profit constitute the sole motivation for cogeneration, the planning process which precedes detailed technical and economic analyses should initially establish if cogeneration is, in fact, a realistic option.

### Fuel and Electricity Costs

Almost any moderately-sized or larger facility which requires coincident supplies of electric power and thermal energy—steam, hot water, chilled water and/or air for drying—is probably a suitable candidate for cogeneration from a technical viewpoint. However, the first and most fundamental factor that establishes if cogeneration is economically feasible is the relationship between the site-specific cost of utility-furnished electricity and the cost of available fuel.

One rule of thumb, which is generally reliable, is that when the first digit of the cost of fuel (expressed in dollars per million Btu) is about half of the cost of utility furnished electricity (expressed in cents per kwh) cogeneration will probably make economic sense. If operational factors favor cogeneration as well, a return on investment in the realm of 20%–30% can generally be anticipated.

### Role of Citizen Groups

Having established the probability that cogeneration will provide the financial benefit anticipated planners must next consider those outside influences which have the potential to impede or deter implementation and to impose substantial costs most often not expected. Often overlooked in the planning process is the possibility that one or more well-intended citizen groups, opposed to anything that even remotely suggests environmental impairment, will successfully delay execution of the project or ultimately prevent its execution. An investigation should be made as to the presence of such a possibility prior to the expenditure of funds.

### Discussion with Local Electric Utility

Similarly, a face-to-face meeting with appropriate representatives of the local electric utility should be included in the planning process. Will they be cooperative or may they be expected to exert every effort to prevent you from cogenerating? Do they have an established Standard Offer to purchase cogenerated electric power under terms and conditions and at prices you believe acceptable, or must you plan on accruing substantial costs incidental to concluding a non-standard negotiated agreement which generally requires utilization of a rate analyst in addition to specialized legal assistance.

If you are planning on a system which will deliver power to the utility, determine how much it will cost to intertie with their distribution network. It is not unusual for a utility to charge a fee of $10,000 to $20,000 to perform the study required in some instances to identify the point in their system where an intertie can be made. It is not unusual for consultants to either ignore the cost of utility intertie in their feasibility analysis, or to underestimate it by orders of magnitude. This cost can amount to millions of dollars in the case of larger systems. Establish what it will be in the planning process.

### Fuel Availability and Cost

If you are contemplating using natural gas as the primary fuel, establish if it will, in fact, be available in the amount required and at what pressure. Will it be necessary for the gas utility to install a new supply line? What will they charge for the line? Will gas be delivered on a Firm or on an Interruptible basis? Gas utility companies are generally willing to furnish a letter commitment in the early planning stages. Get one.

Note that a coal-fired cogeneration system cannot be expected to provide an acceptable return on investment (ROI) when located in an area served by an electric utility which is burning coal as their primary fuel. The economies of scale emphatically mitigate against the possibility of favorable economics in this circumstance.

### Long-Term Fuel Costs

One of the most significant issues which must be addressed in the planning process is the long-term fuel cost scenario upon which ultimate financial viability critically depends. There is no "correct" published fuel cost projection. The recent downward trend in fuel costs was in absolute contradiction to every prior existing projection.

Fuel cost is the largest operating expense. A firm cannot afford to operate a cogeneration system while fuel cost substantially exceeds the original projection, because system viability depends upon revenue derived from the sale of electricity for which it is being paid at a specified level. If a firm is to be paid for cogenerated power at a variable avoided cost and the local electric utility is primarily burning the same fuel the cogenerator intends to burn, ascertain if they have plans to switch to less costly fuel during the economic life of the planned cogeneration system.

There is another side to this coin. Where it is readily evident that cogeneration will pay for itself in very few years, there is no cause to be concerned about the longer term.

### Financing

At an early point in planning, the decision should be made with respect to the method by which the system will be financed.

By investing its own funds, a firm will generally maximize system economics. Leasing obviates the need for capital investment, and concurrently reduces slightly the economic benefits which accrue from cogeneration—by the difference between cost of money and the lessor's charge for the use of his.

Third-party ownership and operation provides the economic benefits of cogeneration with no investment and no financial risk beyond the long-term obligation that a firm will have to "take-or-pay" for prescribed amounts of cogenerated energy at stipulated rates over the term of the contract. As firms offering to install and operate cogeneration systems on a dedicated basis generally assume all up-front costs, it is important to make a decision in regard to this option, if available, prior to spending one's own funds on developing the project.

## Thermal Energy Sales

If state law permits, as it does in New York, and there is an energy-consuming facility proximate to the one for which cogeneration is being evaluated, the planning process should incorporate consideration of serving thermal energy to both entities, or several entities, with a single system. The resulting larger system serving more diversified load requirements has the potential for better economics vis-a-vis a system serving only one facility.

## Data Needed for Feasibility Analysis

The technical and economic feasibility analyses will be based upon the electrical and thermal load profiles for a typical annual operation of the facility to be served by the cogeneration system, and the cost of providing that facility with all forms of energy in its present mode of operation. The planning process should therefore include the accumulation of the facts and data necessary to initiating the detailed analyses. The following should be collected:

- Electric bills for 12 months of typical operation
- Fuel bills for the same period
- Steam recording charts and/or the equivalent recordings for hot water and/or chilled water, showing usage for typical weeks in the four seasons of the year.

- Fifteen minute interval recordings of electric demand, if available.

When recordings of any of these data are not available, personnel responsible for the facility's daily operation should develop an estimate, as accurate as possible, of the hour-by-hour demand for all forms of energy consumed.

### Future Plans

At this juncture it is mandatory to assess future plans as they may affect the capacities of a cogeneration system. Is it likely that the facility will be expanded? Is there a likelihood that production and therefore energy demand will be curtailed in the foreseeable future? Is a third shift contemplated which will incrementally increase energy consumption? Is a product requiring large amounts of energy for its manufacture likely to be dropped from production in the near-term because of perceived changes in the marketplace? Does some significant portion of steam consumption represent leakage from steam or condensate return lines obviously requiring repair or replacement? The influence of the foregoing assessment of energy demand and consumption must be integrated into the data established for the facility's current mode of operation.

### Identify the Optimum System

Now, consideration must be given to the question of who should be given responsibility for the analyses which will identify the type and configuration of the optimum system and establish the basis for commitment, pro or con. If these analyses are not performed by engineers without bias for any of the available options and intimately familiar, as a consequence of directly related successful experience, with operational characteristics and economics of all options, it is unlikely that the optimum system will be identified. The optimum system is the one which will provide maximum savings or maximum ROI on a life cycle basis.

The economic analysis must reflect complete and valid capital cost estimates, as well as service and maintenance costs, reflecting reality rather than wishful thinking, and allowing for deterioration of performance over time, depreciation schedules applicable to different classes of equipment as established by the I.R.S., etc. There is no substitute for experience!

The three types of cogeneration systems which qualify as *bona fide* systems under the definitions established by Federal Energy Regulatory Commission (FERC) Final Rule, Order No. 70 are:

- *Total Energy Systems*—which furnish all required electric and thermal energy to the site completely independent of the electric utility.
- *Selective Energy Systems*—which furnish a selected part of a site's required electricity (the balance furnished by the electric utility) and all or part of the thermal energy requirements. (This type of system is the simplest to identify as being applicable to a facility and has often been incorrectly recommended when a different system would have provided better economics.)
- *Cogeneration Systems*—which either continuously or intermittently deliver electric power to the electric utility—either all electric power produced or that in excess of the site's requirement—while furnishing all or part of the site's thermal energy requirements. A comprehensive analysis of the applicability of this type of system should include consideration of an oversized system which will take advantage of the opportunity, if it exists, to maximize profitable sale of electric power to the utility.

Planning and prefeasibility considerations should reflect the conviction that cogeneration will achieve the predicted savings or earnings established by the detailed feasibility analyses, IF. IF the analyses were realistic and comprehensive, IF the system is reliable and does not experience uncalled-for unscheduled shutdowns, a strong possibility if it is not designed by engineers who have demonstrated their knowledge of the myriad details which influence system reliability. IF all of the system components were designed for continuous duty operation, and have adequate service histories to prove it!

## Technology Considerations

For the most part, cogeneration systems being installed and those that will be installed in the near-term utilize long familiar technology and equipment. While the economic advantages of cogeneration do not demand new technologies, currently there are several commercially available and proven technologies which can enhance system performance and therefore should be considered where appropriate.

- *Direct-Fired and Exhaust-Gas-Fired Absorption Chiller-Heaters,* developed and proven abroad, are now in operation in several systems here in the U.S. These systems will definitely enhance cogeneration system efficiency and economics because of their efficiency which is higher than that of conventional absorption equipment, because they eliminate the need for a hot water or steam boiler, and because they operate at subatmospheric pressure, eliminating the need for licensed operators.
- *Low Temperature Level, Organic Rankine Bottoming Cycle Systems,* which recover shaft horsepower from thermal streams at temperature levels as low as 230°F definitely deserve consideration.
- *High Efficiency, Radial Inflow Steam Turbines,* with efficiency in the realm of 70%, versus 45% to 60% generally expected, can very significantly enhance system economics.
- *Fluidized Bed Boilers* offer the potential capability of making coal use an economic alternative to oil and/or gas use, because they eliminate the need for expensive pollution control equipment by removing sulfur pollutants during the combustion process. Fluidized bed boilers are much smaller than conventional coal-fired steam generators, and can reasonably be expected to be particularly cost effective.

## Benefits of Cogeneration

Attention is particularly directed to the fact that U.S. industry is losing markets to foreign producers who are more energy efficient. The U.S. is 28% less energy efficient than West Germany for example. Cogeneration can reduce the cost of producing product.

When we view the technological progress which has been accomplished, the quantum jumps in efficiency, speed, reliability which science has facilitated in the current century in just about every engineering discipline one can think of, it is a paradox that American government, industry, and business are content to be furnished with utilities provided by systems which differ only slightly from those in use at the turn of the century.

Notwithstanding the recent decline in oil prices and current absence of shortages, many experts believe that long-term fuel availability in the U.S. is a problem, and the need for the energy conservation that cogeneration systems can provide is therefore real, and perhaps more urgent than the present administration's and therefore the general public's perception.

## Concluding Remarks

Cogeneration can have a very important impact on one aspect of our national life that has yet to see very much public attention. This is succinctly summarized in a very detailed report entitled "Energy Policies for Resilience and National Security" dated October 1981, which was presented for the Federal Emergency Management Agency. The summary states, in part:

> The U.S. Energy system is highly vulnerable to large scale failures with catastrophic consequences, and it is becoming more so. We conclude that America's energy insecurity stems from the nature, organization, control structure, and interconnection of highly centralized technologies. These technologies, which unfortunately dominate both the present energy system and current national policy, cannot withstand terrorism, sabotage, enemy attack, natural disaster, or even accidental technical failure. The resulting brittleness of energy supplies poses a grave and growing threat to national security, life and liberty. It also frustrates the efforts of our armed forces to defend a country whose energy supplies can be turned off by a handful of people.

# Chapter 3

# Technical and Economic Feasibility Evaluation

*Dilip R. Limaye*

## Technical Options for Cogeneration

Cogeneration can be achieved by "topping" or "bottoming" cycles. Topping cycles involve the secondary utilization of thermal energy after the electricity generation process. (In some cases, the thermal energy would have been conventionally treated as "reject heat" and have no value.) In bottoming cycles, on the other hand, thermal energy is used in an industrial process first, and the energy which would normally be rejected is used to generate electricity.

A number of different options are available for topping cycles. These include:

- Extraction steam turbines
- Back-pressure steam turbines
- Gas turbines
- Gas turbines with waste heat boiler
- Combined cycles (steam turbine and gas turbine)
- Diesel and gas engines
- Fuel cells
- Other new technologies.

Bottoming options include:

- Low pressure Rankine cycles

- Stirling cycles
- Brayton cycles.

Most existing cogeneration systems use steam turbines (extraction or back-pressure), gas turbines or diesels. Steam turbines, of course, represent the most prevalent method for electric power generation. For cogeneration, steam is taken from the turbine at a pressure and temperature appropriate for the process energy needs (generally much higher than the energy conventionally rejected from a power plant).

This is achieved by extracting the steam at an intermediate step in the turbine (extraction turbine) or by having the steam exhausted from the turbine at a high pressure (back-pressure turbine). The result is a decrease in the amount of electricity produced per unit of steam and an increase in the availability of thermal energy.

Gas turbines are also conventionally used for power generation. The exhaust from a gas turbine can be used as hot air for process use or passed through a waste heat boiler to generate steam. For a given quality of steam requirements, gas turbines can produce more electricity than steam turbines. However, under present technology, gas turbines need natural gas or distillate oils as input fuels, while steam turbines (at least large installations) can use coal-fired boilers.

Diesel engines have a higher conversion efficiency than gas turbines but also require petroleum-based fuels. Steam turbine systems are generally economically feasible only in large sizes (over 10 mw). Gas turbines can be used to intermediate or large sizes—there are many in the 1–10 mw range. Diesels can be as small as 100 kw.

Combined cycle cogeneration or new technologies such as fuel cells with heat recovery are likely to be attractive technical options because of the possibility of decoupling the electric and thermal outputs (changing the ratio of electric and thermal output). Other new technologies, including solar and geothermal, can also be used to generate electricity and thermal energy, and are currently being researched, but are not likely to achieve significant penetration in the 1980s.

Bottoming applications depend on the quality (temperature and pressure) of reject heat from an industrial process. Low-pressure steam turbines can be used with reject heat temperatures of 400°F to 1000°F. The electrical efficiency is, however, low. Organic Rankine cycles which use a process similar to steam turbines, but with organic fluids, can be used with reject heat streams as low as 150°F. With high temperature boiler and furnace exhausts (450°F), Stirling cycles can also be used, and with very high-temperature streams, Brayton cycles can be employed. The potential for bottoming cycle cogeneration appears to be limited in the 1980s.

Table 3-1 shows some of the technical characteristics of cogeneration systems.

## The Economics of Cogeneration

The changing economics of energy have made cogeneration an attractive option for industry. Currently available and emerging technological options can be used to provide industry's thermal needs and generate power for the utility grid. Also, as discussed in Chapter 1, Federal legislation has attempted to remove most of the institutional barriers to industrial cogeneration. State implementation of the Federal rules, expected shortly, will allow industries to cogenerate without fear of utility-type regulation, and obtain a reasonable price for exports of electricity. The legislation also prevents high standby charges. However, a careful evaluation of cogeneration economics must be performed before investing significant capital. A number of analytical tools are available to perform such economic evaluation.

It is important to note that the economic evaluation of cogeneration must adequately consider utility perspectives and roles. Since the price paid by the utility for purchase of power from the industry is based on the avoided cost, which depends on the generation mix, fuel types and costs, and anticipated capacity expansion, the changing economics of the utility's generation are important to the cogenerator. The perspective of the utility must therefore be understood by the cogenerator, and included in his economic analysis.

**Table 3-1. Cogeneration Topping Cycle Performance Parameters**

| *Cogeneration Systems* | *Electrical Capacity of a Single Unit (kw)* | *Heat Rate*[2] *(Btu/kwh)* | *Electrical Efficiency (%)* | *Thermal Efficiency (%)* | *Total Efficiency (%)* | *Exhaust Temperature °F* | *Steam lbs/hr Generation @ 125 psig* |
|---|---|---|---|---|---|---|---|
| Small reciprocating Gas Engines | 1–500 | 25,000 to 10,000 | 14–34 | 52 | 66–86 | 600–1200 | 0–200[1] |
| Large reciprocating Gas Engines | 500–17,000 | 13,000 to 9,500 | 26–36 | 52 | 78–88 | 600–1200 | 100–10,000[1] |
| Diesel Engines | 100–1,000 | 15,000 to 11,000 | 23–31 | 44 | 67–75 | 700–1500 | 100–400[1] |
| Industrial Gas Turbines | 800–10,000 | 14,000 to 11,000 | 24–31 | 50 | 74–81 | 800–1000 | 3,000 to 30,000 |
| Utility Size Gas Turbines | 10,000–75,000 | 13,000 to 11,000 | 26–31 | 50 | 76–81 | 700 | 30,000 to 300,000 |
| Steam Cycles | 5,000–100,000 | 50,000 to 10,000 | 7–34 | 28 | 35–62 | 350–1000 | 10,000 to 100,000 |

[1] Hot water @ 250°F is available at 10 times the flow of the steam.

[2] Heat rate is the heating value input to the cycle per kwh of electrical output. The electrical generation efficiency in percent of a prime mover can be determined from its heat rate by the following formula:

$$\text{Efficiency} = \frac{3413}{\text{Heat Rate}} \times 100$$

## The EPRI Project

In a current project sponsored by the Electric Power Research Institute (EPRI) to evaluate cogeneration alternatives, Synergic Resources Corporation is developing computerized evaluation tools to assess the costs and benefits of cogeneration.[1] The objectives of the EPRI project, called "Evaluation of Dual Energy Use Systems (DEUS) Applications" are to:[2]

- Develop a methodology to assess cogeneration options, with explicit consideration of utility perspectives and impacts.
- Identify promising candidate applications for cogeneration.
- Identify and assess utility options for participation in industrial cogeneration.
- Identify research, development and demonstration needs and priorities.

The first step in this study was to conduct surveys and case studies of existing cogeneration facilities to identify the site-specific factors which influence successful implementation of cogeneration. A methodology for screening and evaluation of cogeneration applications has been developed and is described in a recent paper.[3] The methodology will be supported by a data base on the performance and cost characteristics of existing cogeneration facilities.

## Methodology for Cogeneration Evaluation

The methodology consists of two steps. In the first step, the aggregate benefits, costs and impacts on the utility, industry and society. This calculation is based on the value of electric and thermal energy used, the costs of producing these outputs, and the related social and environmental considerations.

Institutional and regulatory considerations such as standby and buy-back rates (PURPA rates), tax credits, alternative arrangements for ownership and operation, etc., do not affect the overall benefits of cogeneration from the systems viewpoint, but do determine how the benefits, costs and impacts are shared by the various affected parties. Such institutional and regulatory factors are therefore considered in the second step under each type of arrangement for ownership or operation. These considerations influence the negotiated position of each party relative to the cogeneration venture.

An overview of the first step is shown in Figure 3-1. Using information regarding the characteristics of cogeneration technologies, the energy needs for the application, and local utility data, the size of the cogeneration system is determined under alternative sizing options. Calculations are then performed for the performance of the cogeneration system and its capital, operating and maintenance costs.

The performance calculations provide information regarding the amount of thermal and electric energy generated by the cogeneration system under different operating strategies. The value of the power generated is then calculated based on data on the utility's generation mix and expansion plan.

Similarly, the value of thermal energy generated is calculated based on the alternative costs of thermal energy generation for the industry. An economic analysis is then performed, taking into account the value of the thermal and electric outputs relative to the capital and O&M costs under each sizing and operating option.

The economic data are then compared to the conventional energy generation systems to determine the aggregate costs, benefits and impacts of the cogeneration option. By performing these sets of calculations for different cogeneration technologies and different sizing and performance options, the most attractive options can be identified.

Figure 3-2 shows an overview of the second step, the detailed analysis of the cogeneration options. For each option considered to be an attractive option, an analysis of the institutional and regulatory constraints is performed. Based on this analysis, the alternative organizational and financial options are identified. For each of these options, an analysis of the impacts on the utility and industry is then performed.

Where appropriate, if third party considerations are important, the analysis includes the impact on such third parties. In this step, a detailed evaluation of the economic, financial and regulatory aspects is performed from the point of view of the utility and industry to provide information regarding the alternative methods of allocating the benefits of the cogeneration option.

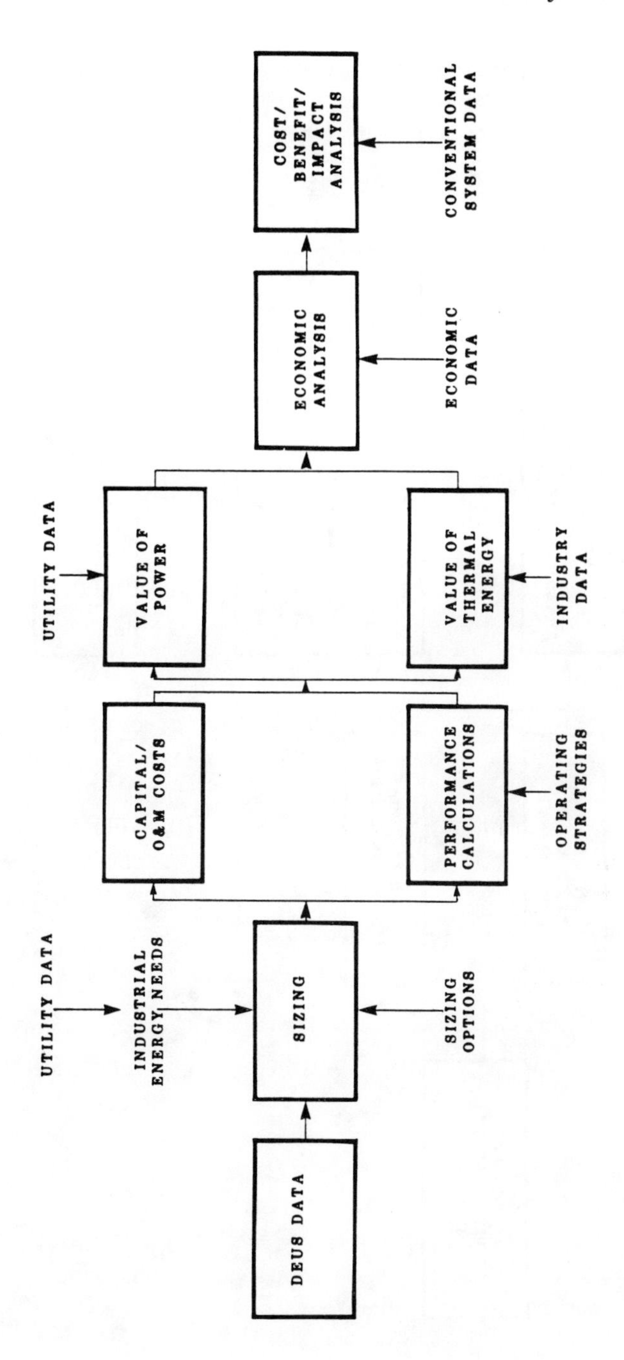


Figure 3-1. Suggested Approach for Evaluation of DEUS Options
Step 1 — Identification of Attractive Options

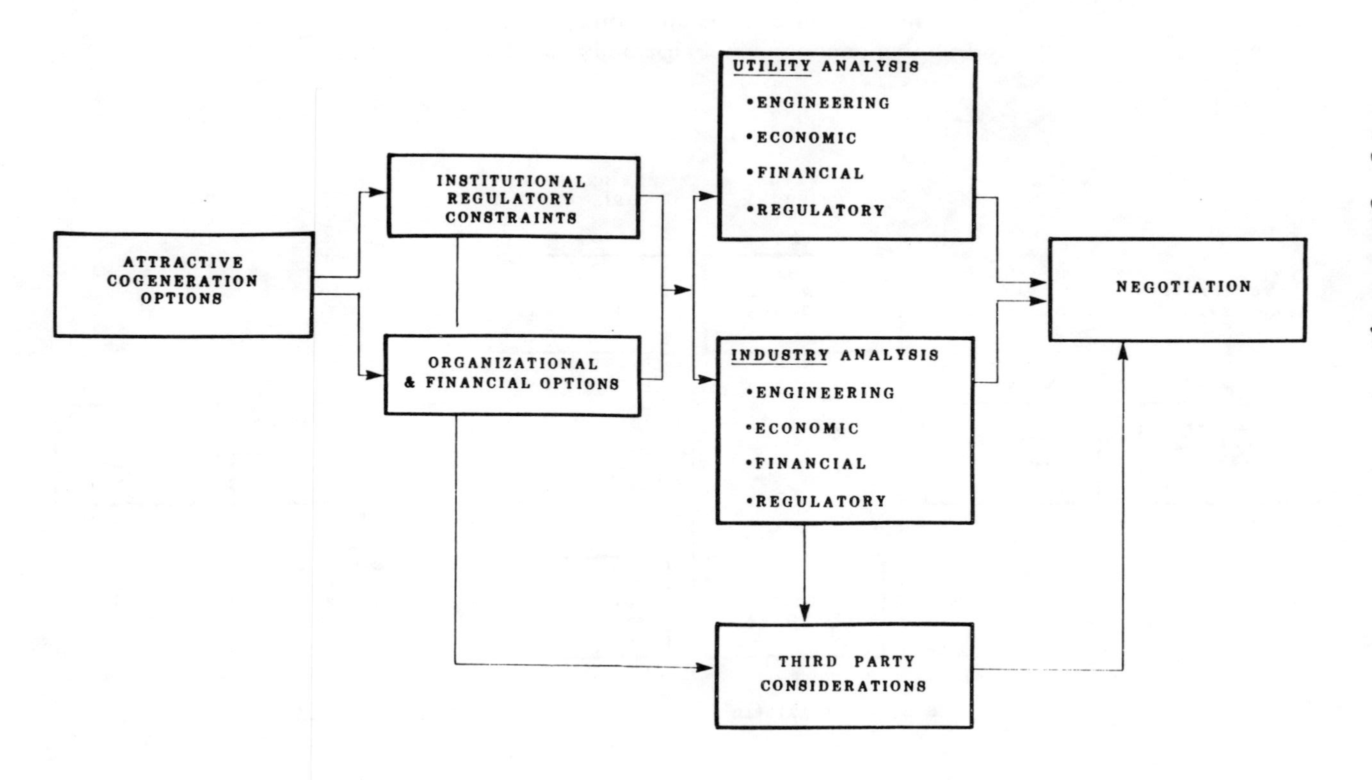


Figure 3-2. Suggested Approach
Step 2 – Detailed Analysis of Options

It is hoped that this analysis will provide all concerned parties with adequate information to enter into a meaningful negotiation process which will lead to the implementation of the most attractive cogeneration systems.

## Value of Thermal and Electric Energy

The value of thermal energy produced by a cogeneration system can be calculated as equal to the costs of alternative generation of such energy in a conventional plant, taking into account the plant's requirements for thermal energy supply reliability. In order to calculate this value, it is necessary to determine the fuel costs at the site, and the costs of installing a boiler or other means of generating the required thermal energy. The operating hours of the plant, thermal load factors and other operating characteristics will have to be considered in determining these costs.

The value of the electric power generated by the cogeneration system consists of two parts—the energy value and the capacity value. The energy value of the power can be calculated taking into account the following considerations:

- The amount and the type of fuel saved by the local electric utility because of the availability of cogenerated power;
- The variation of the available power by time of day and the related fuel used by time of day for the utility;
- The variation of fuel use and power generated by season, if any;
- The future changes in the fuel mix and fuel prices, expected over the lifetime of the cogeneration facility;
- Any savings in operating and maintenance costs for utility plants;
- Possible reductions in transmission and distribution losses for the utility system.

The capacity value of the power generated also depends on a large number of considerations. These include the following:

• *The availability of the power to the utility*—In order to realize credits for capacity, the power generated by the cogeneration system must displace utility capacity over some period of time. If the cogenerated power is not available when the utility needs it, then the utility will have to back up the cogeneration system with additional capacity. In such situations, the capacity credit would be very small, or nonexistent. On the other hand, if the cogenerated power is available at all times when the utility needs it, then there should be some capacity credit given to the cogeneration system.

• *Reliability*—While no power generation facility is likely to be 100% reliable, experience with cogeneration facilities shows that they can accomplish a high degree of reliability with a small amount of unscheduled maintenance. In general, the higher the reliability of the system, the greater should be its capacity value. Some utilities have argued that they would have to back up cogeneration facilities with enough standby capacity and that there would be no avoided capacity costs. However, if the reliability of the cogeneration system is adequately accounted for in the utility's calculations of loss of load probability and reserve margin, then an appropriate method can be determined for developing the proper capacity value.

• *Long-term availability of power*—In many cases, the capacity value of a cogeneration system will have to be calculated based on displaced utility capacity over some future planning horizon. This requires some guarantees of the long-term availability of power from the cogenerator. In general, a cogenerator prepared to guarantee long-term availability through a long-term contract is likely to have a greater value for capacity than one with some uncertainty regarding the long-term availability of power.

• *Supply diversity*—Given a number of cogenerators on a utility system, the supply diversity or the probability of outages of one or more cogenerators should be calculated in determining the appropriate capacity credits. This can be accomplished by treating each cogenerator as another unit in the utility system available to meet the utility's loads. The characteristics of power output, forced outage rates and maintenance schedules for each cogenerator can be analyzed using the utility's evaluation methodologies. The greater the diversity of supply, the greater the capacity value of the cogenerator.

• *Short-term versus long-term considerations*—Many utilities with excess generation capacity have argued that they should not provide any capacity value to potential cogenerators. Their arguments are probably valid in the short-run. If utilities do not save any capacity costs by having cogenerated power available, then the short-term capacity credit should be zero. Short-term capacity credits are relevant only for utilities with low current reserve margins, or utilities with substantial purchased capacity. In the long run, however, the situation is different. If a utility has excess capacity now, but is experiencing some load growth, it may have to add capacity in the future. The availability of the cogenerator will allow such capacity additions to be either deferred or cancelled, leading to some savings in investment costs. Such savings should be reflected in the development of the capacity value of the cogenerator.

• *Other factors affecting generation capacity credits*—Other factors which influence the capacity value of a cogenerator include the quality of the power generated, the degree of operating control that the utility has over a cogeneration system, the size of the cogenerator, and the possible value of the cogenerator for spinning reserve.

• *Transmission and distribution capacity credits*—It is possible that a cogeneration system would reduce the need for transmission and distribution capacity additions. The calculation of avoided transmission and distribution (T&D) capacity has to be site-specific and is extremely difficult. It requires the analysis of the reliability of supply at the customer level, which includes an assessment of the reliability of the T&D network. If the cogenerator is sufficiently large and is located near a load center, it is possible that it could lead to the deferral or elimination of some future T&D investments by the utility. In such cases, the cogenerator should be given an appropriate capacity credit.

## Computer Models for Feasibility Analysis

In the EPRI project mentioned above, two computer models have been developed to perform technical and economic feasibility analysis of cogeneration projects.

- *Computer Evaluation of Dual Energy Use Systems (DEUS)*—In order to perform the sizing and performance calculations, and to screen and evaluate the costs and benefits of cogeneration options relative to a conventional system, an analytical model called DEUS—Computer Evaluation of Dual Energy Use Systems, has been developed.[4] This model accomplishes step 1 of the evaluation methodology. An overview of this model is provided in Figure 3-3. The model can evaluate up to twelve systems (including a no-cogeneration base case), taking into account industrial requirements for heat and power, fuel types, utility rate schedules, (including industrial and PURPA rates), economic data, operational ground rules, and various ownership types.

In many industrial processes, the actual process thermal and power demands vary with time of day and/or seasonally. To be compatible with anticipated PURPA rate schedules, the program has the capability to represent 36 time periods per year. For example, the 36 time periods might be used to cover four seasons, three types of days per week, and three time periods per day (on-peak, near-peak, and off-peak). The program has the capability to evaluate DEUS configurations incorporating up to four fuel streams, with each fueling a given type energy conversion system (ECS).

- *COPE—Cogeneration Options Evaluation*—A computer model called COPE—Cogeneration Options Evaluation, has been developed to calculate after tax cash flows to the utility, industry and, where appropriate, third parties.[5] COPE can handle all practical ownership and financial arrangements and account for tax credits, depreciation and other relevant financial and economic parameters, taking into account the most recent legislation and regulations. COPE is designed to provide information to all potential participants in a cogeneration venture so as to identify mutually beneficial institutional arrangements (see Figure 3-4).

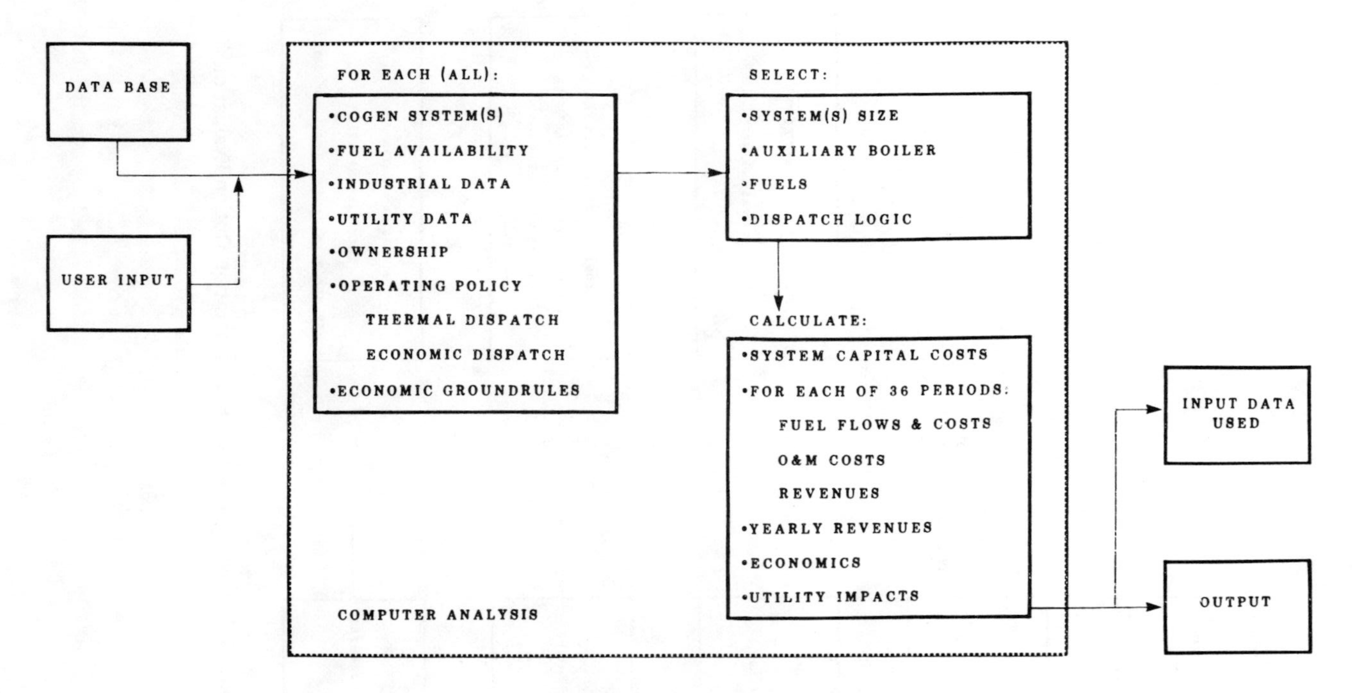


**Figure 3-3. Overall Structure of DEUS Program**

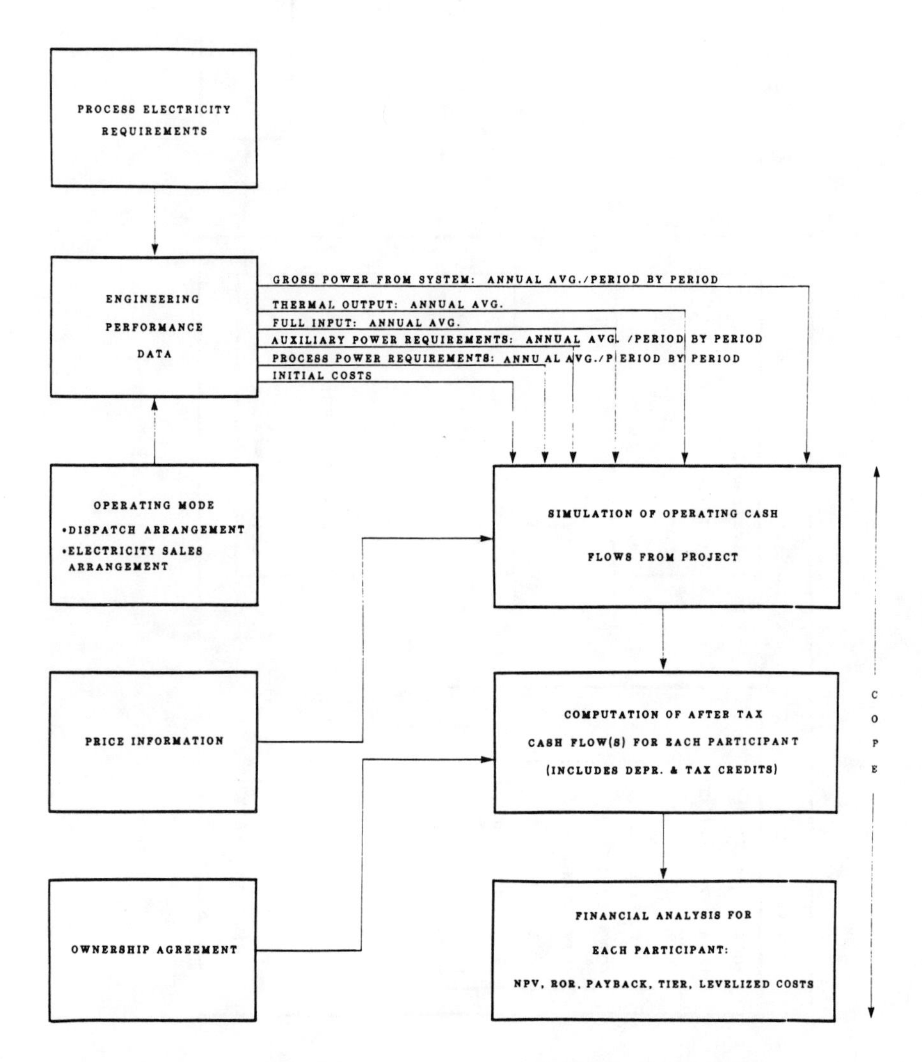


Figure 3-4. Overview of COPE Cogeneration Option Evaluation

The magnitude and distribution of after-tax costs and benefits of a cogeneration system are significantly influenced by its ownership structure (utility, industry, third party), operating mode (thermal dispatch versus utility economic dispatch) and the electricity sales arrangement (simultaneous buy-sell, buy-shortage/sell-excess). COPE is designed to evaluate alternative combinations of ownership, operating modes and sales arrangements.

In the past, a common assumption was that a cogeneration system is owned entirely either by an industry or a utility. With the increased interest in cogeneration, a number of innovative arrangements are being considered. For example, joint ventures among industry, utility and third parties may offer benefits to all the participants. One arrangement to form a joint venture is to create a separate corporation for the sole purpose of owning and operating the cogeneration project. In this arrangement, the cogeneration project would be taxed as a corporation.

The partnership arrangement can also be used to form joint ventures. Partnerships do not pay a Federal tax on earnings comparable to the corporate earnings tax; however, each partner pays Federal tax on his share of earnings from the partnership. Also, partnerships enjoy considerable flexibility in the apportionment of tax and depreciation benefits as well as profits (or losses) among partners. It is possible, therefore, to design partnership arrangements so as to attract private (third party) investors by offering them substantial tax-related benefits. At the same time, third parties, having no site-specific thermal or electric requirements, are unlikely to insist on specific operating modes. Thus, partnerships between utilities, industries and third parties could often by mutually beneficial.

COPE is designed to analyze any one of the following ownership arrangements. The utility can be either an investor-owned or a tax-exempt utility.

- 100% Ownership
  - —100% Utility Ownership
  - —100% Industry Ownership
  - —100% Third Party Ownership (or Separate Corporation).

- Joint Ventures
  —Partnership - Utility/Industry
  —Partnership - Utility/Third Party
  —Partnership - Industry/Third Party
  —Partnership - Utility/Industry/Third Party.
- Leasing Arrangements
  —Lessor/Lessee - Third Party/Utility
  —Lessor/Lessee - Third Party/Industry.

*Illustrative Results*—Illustrative results of the application of these models for the economic evaluation of cogeneration in a pulp mill are shown in Figure 3-5. The results indicate that a 59 mw cogeneration system offers a 25% rate of return on incremental investment over a no-cogeneration case. A 100 mw cogeneration system offers a 15.6% rate of return. The revenues from electricity sales in the 100 mw case are comparable to income from pulp sales. Figure 3-6 shows the rate of return vs. size for the pulp mill application.

| | *No Generation* | *Assumed Thermal Match* | *Maximum Cogeneration* |
|---|---|---|---|
| Gross MW Output | 0 | 59.1 | 100 |
| Total Installed Cost of Power Plant (Million $) | 88 | 146 | 202 |
| Cost chargeable to Power Plant Generation (Million $) | – | 58 | 224 |
| Annual Operating & Maintenance Costs (Million $) | 2.98 | 6.09 | 7.01 |
| Annual Fuel Costs (Million $) | 7.29 | 17.48 | 33.10 |
| Annual Cost of Purchased Electricity (Million $) | 13.04 | 13.91 | 15.02 |
| Annual Electric Revenues (Million $) | 0 | 37.71 | 61.07 |
| Projected Return on Investment (%) | 0 | 25.4 | 15.6 |

**Figure 3-5. Illustration of Cogeneration Economics Pulp Mill Example**

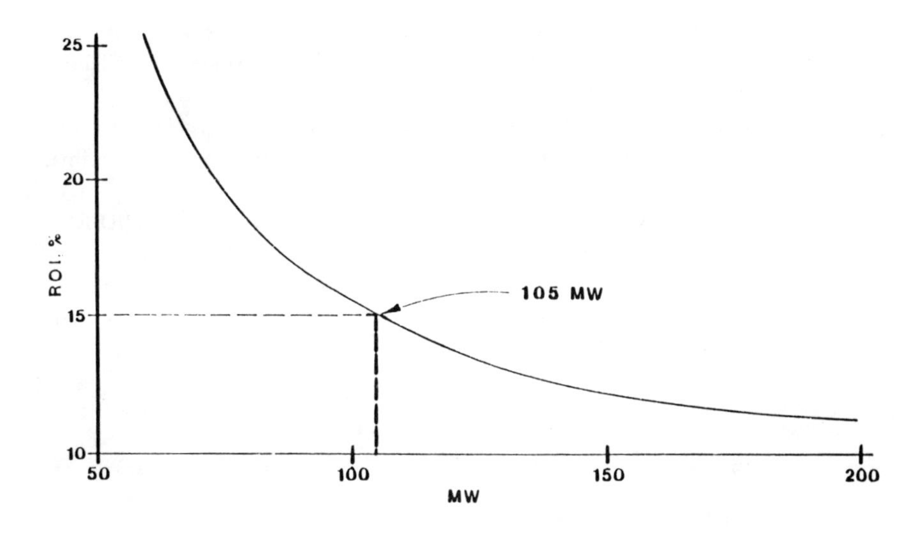


**Figure 3-6. Projected Return on Investment vs Megawatt Size 1985 Conceptual Design, 1000 Tons/Day Bleached Kraft Pulp Mill West Coast, U.S.A.**

## Acknowledgment

This paper is partly based on the author's research sponsored by the Electric Power Research Institute (EPRI) under Project RP1276, Evaluation of Dual Energy Use Systems (DEUS). The assistance and cooperation of Robert Mauro and Dr. S. David Hu of EPRI are gratefully acknowledged.

## References

1. Synergic Resources Corporation, *Evaluation of Dual Energy Use Systems: Volume I, Executive Summary*, EPRI EM-2695-SY, Electric Power Research Institute, Palo Alto, CA, 94303.

2. Synergic Resources Corporation, *Evaluation of Dual Energy Use Systems (DEUS) Applications: Project Overview*, RP1276, prepared for Electric Power Research Institute, August 1980.
3. Dilip R. Limaye, *Methodology for Evaluation of Cogeneration Projects*, paper presented at the National Fuel Cell Seminar, Norfolk, Virginia, June 1981.
4. General Electric Company, *DEUS Computer Evaluation Model*, EPRI EM-2776, Electric Power Research Institute, December 1982.
5. B. Vankateshwara and D. R. Limaye, *COPE—Cogeneration Options Evaluation; Program Descriptive Manual*, Electric Power Research Institute, Palo Alto, CA, 94303.

# Chapter 4

# Assessing Cogeneration Feasibility

**Harry Davitian**

In planning a cogeneration system, choices must be made regarding various design options. Depending upon the type of facility in which the cogeneration system is to be placed, decisions may have to be made regarding the prime mover, system configuration, capacity, operating mode, whether or not power is to be sold to the utility, and whether backup is to be provided by the utility or independently. This chapter describes the process by which such decisions can be made. To add depth to the discussion, the process is considered in the context of a specific application and a specific region—cogeneration in large buildings located in New York City. Buildings of the type we will be considering can accommodate a cogeneration system of approximately 1-3 mw or more.

At present, over 112 mw of cogeneration are installed in 19 facilities in New York City and as much as 1500 mw of additional capacity could be added in the future under favorable economic and regulatory conditions. Electricity rates in New York City are high—equivalent to 12 to 15¢/kwh—providing a strong inducement to seek alternatives to utility electricity. The State has a minimum purchase rate of 6¢/kwh for power sold to the electric utility.

In addition, a number of the buildings in the City are cooled by absorption chillers, facilitating conversion to cogeneration. On the other hand, a typical large building cannot support a cogeneration system larger than 1-2 mw, a size range in which economies of scale are not fully realized, and standby rates and the cost of interconnection in New York City are high.

Because of the complex interplay among these various positive and negative factors, a careful analysis of specific facility load patterns and cogeneration system configurations and operating modes is necessary to develop a realistic assessment of economic feasibility.

## Interchanges with Utility—Selling Power and Buying Backup

Frequently, the decisions regarding whether to generate and sell excess power to the utility and whether to purchase backup from the utility can be made independently of the consideration of cogeneration system configurations and operating modes. This is possible when the choice of cogeneration technology is not dependent on whether power is sold or what the source of backup power is. If the technology choice is not so dependent, then an initial determination can be made regarding whether excess power will be generated and sold and how standby power will be handled, greatly simplifying the overall feasibility analysis.

For building cogeneration systems in the 1–3 mw size range, technological options are quite limited; generally, only natural gas or distillate fueled reciprocating or combustion turbine systems are realistic possibilities. Hence, it is possible to make an initial determination as to whether or not to sell power to the utility and how standby power should be provided. In this section, we demonstrate how this is done.

### *Sales of Power to the Utility*

Since utility electric rates in New York are 12–15¢/kwh, while the rates for purchase under PURPA are only 6¢/kwh, a cogenerator would never engage in arbitrage (i.e., selling all generated to the utility while simultaneously purchasing all the power consumed from the utility) but would only sell the excess of its generation over its consumption.

For cogenerators wishing to sell power, the added interconnection equipment needed to export excess power can be expensive. For a large facility, the minimum cost of the additional interconnection equipment in a dense urban area is likely to be approximately $100,-000. This covers the cost of transformers, protection equipment, control systems, metering, and modifications to other interconnection points with the utility.

Unless a utility primary feeder already goes directly to the facility site, it will also be necessary to install a primary line to interconnect with the electric utility's nearest substation bus or to a nearby primary feeder. Cable costs may range from $25,000 to $100,000. In addition to the installation of the cable itself, auxiliary control systems and protection equipment are required. In combination, the primary line and the associated auxiliary equipment will cost a total of approximately $300,000.

The question is: Under what conditions will the income from sales of excess power be large enough to justify the investment in the additional equipment?

The key variables that affect the profitability of power sales are (1) whether a primary feeder already comes directly to the facility, (2) whether additional cogeneration capacity is required for the purpose of selling power, (3) the utility's purchase rate, (4) the value of the thermal energy produced by the cogeneration system as measured by the cost of the thermal energy source that is being displaced, and (5) the power level at which the cogenerator wishes to sell power.

We evaluated the economic feasibility of selling power from a natural gas or distillate fueled reciprocating engine cogeneration system under a variety of circumstances (see Table 4-1). To determine whether a cogenerator should make the investment in the additional equipment needed to sell power, a simple payback period calculation was performed based on a single year analysis; i.e., the net annual operating income generated by selling power to the utility was compared to the total incremental capital investment required to sell power. The assumptions used in this calculation were:

Cost of fuel . . . . . . . . . . . . . . . . . . . . . . . . $6 million Btu
Minimum interconnection cost . . . . . . . . . . . . . . $100,000
Cost of additional cogeneration capacity . . . . . . $700/kw
Efficiency of boiler displaced by
cogenerated thermal energy . . . . . . . . . . . . . . . . . . 70%

Ratio of cogenerated thermal to electric output . . . . 1.3:1
Gross heat rate of cogeneration system . 12,000 Btu/kwh
O&M for cogeneration system . . . . . . . . . . . . . . . 1¢/kwh
Utility's rate for purchase of power . . . . . . . . . . . 6¢/kwh
Utilization of recovered thermal energy . . . . . . . . . . 100%

In none of the cases examined was it worthwhile for a cogenerator to purchase additional cogeneration capacity simply in order to sell power; the payback periods under these conditions were over seven years. This is true even if the additional equipment operates at a high load factor and virtually all of the thermal energy is utilized. Even if additional cogeneration capacity does not have to be purchased, cogenerators selling less than about 1 mw of power at a moderate to high capacity factor (i.e., greater than about 40%) will be unlikely to find an investment in interconnection equipment feasible. Only if the cogenerator is already hooked up to the utility through a primary line and if no additional investment in cogenerating capacity is required does the payback for the investment in additional equipment fall in an attractive range.

Because a portion of the interconnection costs are fixed and independent of the amount of power sold, payback periods are considerably reduced as power sales increase to 3 mw. Even if a primary line must be installed, payback is under three years. Other cases examined (not shown in Table 4-1) show that as the cost of fuel rises from $6 to $7 per million Btu, the payback becomes unattractive in almost all cases.

In summary, the conditions requisite to a profitable investment in interconnection equipment for the purpose of selling power to the electric utility are (1) the ability to sell approximately 1 mw of power or more at a moderate to high capacity factor, (2) a high fractional utilization of the thermal energy recovered when generating power for sale, and (3) fuel prices of about $6 million Btu or less. Also, for facilities selling significantly less than 3 mw, the preexistence of a primary feeder linking the facility to the electric utility is necessary to make sales of power worthwhile.

**Table 4-1. Payback Associated with Investment in Equipment Required to Sell Power to Utility**
*(All $ figures in 1000's)*

| *Primary Line Cost* | *Power Sold (mw)* | *Incremental Investment to Sell Power* | *Annual Operating Savings* | *Payback (Years)* |
|---|---|---|---|---|
| **Case A: No Incremental Cogeneration Capacity Cost; Load Factor of Power Sold = 0.4** | | | | |
| 0 | 1 | 100 | 53 | 1.9 |
| 0 | 3 | 100 | 159 | 0.6 |
| 300 | 1 | 400 | 53 | 7.5 |
| 300 | 3 | 400 | 159 | 2.5 |
| **Case B: Incremental Cogeneration Capacity Cost = $700/kw; Load Factor of Power Sold = 0.8** | | | | |
| 0 | 1 | 800 | 106 | 7.5 |
| 0 | 3 | 2200 | 318 | 6.9 |
| 300 | 1 | 1100 | 106 | 10.4 |
| 300 | 3 | 2500 | 318 | 7.9 |

### *Utility vs. Independent Backup*

Should backup power be purchased from the utility or should additional generating equipment be installed to supply independent standby capacity?

To compare these alternatives, it is assumed that 1 mw of standby capacity is required and that inexpensive generating capacity in the form of a high speed reciprocating engine-generator set can be installed for a total cost of approximately $400/kw. It is further assumed that the investment is financed through a self-amortizing, 12% interest, 10-year loan.

A useful parameter for comparative evaluation is the minimum annual cost of standby power to the facility owner assuming no breakdowns occur during the year. Con Edison's Backup tariff consists of an energy charge and a relatively high, ratcheted demand charge.* The demand charge is composed of two components: a contract demand charge to be paid independently of actual usage, and an "as used" charge proportional to the actual demand applicable during the summer peak period. The peak period demand charge is ratcheted for a year but applies only to the summer months.

Under Con Edison's Backup tariff, the minimum annual cost to a customer connected at the primary voltage level is about $44,000. Similarly, a secondary customer would incur costs of about $79,000. The comparable cost for independent backup is $71,000 (see Table 4-2).

**Table 4-2. Comparison of Annual Costs of Utility Backup with Costs of Self Backup**
*(Units: 1000's of 1982 $)*

| | | |
|---|---|---|
| **Utility Backup** | | |
| | *Primary* | *Secondary* |
| Contract demand charge | 44 | 79 |
| As used demand charge—summer peak outage | 36 | 71 |
| Minimum annual cost—no summer outage | 44 | 79 |
| Minimum annual cost—with summer outage | 80 | 150 |
| **Independent Backup** | | |
| Installed Cost of Back-up Generator | 400 | |
| Annual cost with 10 year loan @ 12% | 71 | |
| Minimum annual cost | 71 | |

However, a cogenerator must consider the possibility that the system will suffer an outage and that the outage may occur during the summer peak. Should that happen, the total annual demand charges for standby capacity at the primary level is about $80,000 (i.e., the

*The New York Public Service Commission has decided to permit cogenerators to purchase power for standby and supplementary use under the normal tariff that would otherwise apply to the facility. These tariffs do not include ratchets. An examination of this option is not included here.

total annual cost incurred upon use of the very first kwh of backup during the summer peak), or $150,000 if the customer is connected at the secondary level. Of course, the comparable minimum cost of independent backup is still $71,000/year.

## Economic Feasibility Analysis

### *Analytical Approach*

Because the cost of storing electrical and thermal energy is relatively high, the detailed dynamic characteristics of the thermal and electrical loads of the facility play an important role in the economics of cogeneration. The cost effectiveness of cogeneration is also affected by the variation of the electric tariff by time of day, week and year.

In examining the feasibility of cogeneration, it is important that the effects of these independent dynamic variables (i.e., electric loads, thermal loads, and electric tariff) be treated accurately. We use an hourly simulation model—the DOE-2 model developed at Lawrence Berkeley Laboratory—to simulate the thermal and electrical loads of the building and the performance of a conventional heating, ventilating, and air conditioning (HVAC) system and of a cogeneration system in that building.

The model's inputs include data on the building's structure, occupancy pattern, equipment usage, and HVAC system as well as historical hourly weather data. The model can simulate the performance of a cogeneration system under various operating modes (e.g., baseload, thermal load following, and electrical load following). The primary output of the cogeneration simulation portion of the model is the system's electrical and thermal output and fuel consumption.

For the New York City study, several dozen DOE-2 cases were run to examine the performance of various cogeneration system configurations and operating strategies in different types of buildings. Five building types—multi-family residential, commercial, hospital, hotel, and university—were modeled based upon their prevalence in the urban landscape.

DOE-2 has an economic evaluation routine but it is limited, especially in that it can treat only relatively simple electric tariffs. To compute annual payments under the more complicated time-of-use demand and energy charge structures applicable in New York, we used our RATES model which can treat virtually any type of utility tariff.

For the New York study, utility payments for 1982 were computed from the hourly electric load output of the DOE-2 model and payments for future years were derived by multiplying the 1982 result by the relative change in utility average costs between the two years as given by a utility cost analysis performed using our ENTEK-1 model.

Annual operating savings and the net present value and internal rate of return associated with cogeneration investments were computed using our INVEST model. This model computes both before-tax and after-tax investment indices.

*Technical and Economic Assumptions*

*System Type and Configuration.* The cogeneration systems were assumed to use natural gas-fired reciprocating engines because of the relatively low intial costs and the high overall performance of this engine-fuel combination in the size range of interest. The installed cost was assumed to be $800/kw (1982 $) including basic heat recovery equipment. Auxiliary generators at a cost of $400/kw were assumed to be used for standby. No power was assumed to be sold to the utility. However, where needed, a standard utility intertie was assumed for the purpose of providing supplementary power (power used regularly to supplement the output of a cogeneration system not designed to meet the full electrical requirements of the facility).

*Fuel and Electricity Prices.* Fuel prices were assumed to escalate at 2%/year (real). Starting from a 1982 level of $5.70 million Btu, electricity prices were assumed to decrease slightly in real terms over the 20-year study period in accord with the utility system cost analysis we performed as part of the study.

*Operating Pattern.* In the economically preferred operating mode, the cogeneration system follows the electrical load (thermal load following was tested for several cases but the operating cost savings were significantly lower), and the cogeneration system output displaces purchases from the utility with no power being sent to the grid.

*Effect of System Size on Investment Economics*

Varying the cogeneration capacity at a facility affects the economics in several ways: As the size is increased, economies of scale occur because a major portion of the costs of interconnection and standby power are constant and do not increase with system size. Increasing the system size also reduces demand charges by a corresponding amount and reduces the need for expensive utility power during peak periods.

On the other hand, a larger system wastes more thermal energy and has a lower capacity factor; both reduce the cost effectiveness of the system. For each facility, an optimum point was found where these effects balanced (see Figure 4-1).

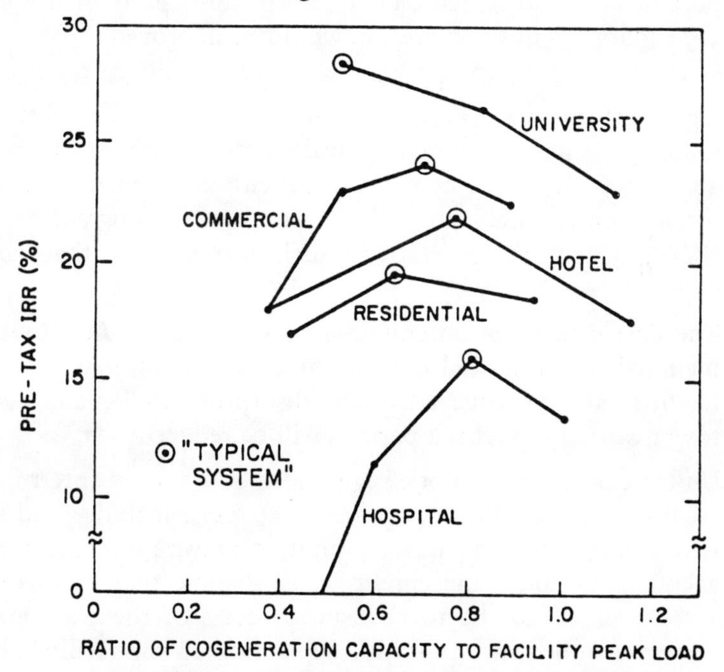


**Figure 4-1. Relationship Between Return on Investment and the Relative Capacity of the Cogeneration System**

### *Return on Cogeneration Investment for Typical Facilities*

For the optimal cogeneration system size and operating mode, payback periods range from three to six years, and pre-tax internal rates of return range from 13% to 28% (see Table 4-3). For the residential, commercial, and hospital facilities representing the bulk of the potential market, the pretax IRR is under 20% and the after-tax IRR under 13%.

In assessing the broader implications of these results to other utilities and urban areas, it is important to consider the assumptions that underlie the investment analysis. Key economic, facility, and ownership related assumptions, which may not be applicable in other regions or to specific actual facilities, are:

- The fuel was assumed to be natural gas with a cost of $6/million Btu, increasing at 2%/year. To the extent that fuel can be obtained at a lower cost (as it can in some parts of the U.S.), the economics of cogeneration would be improved.
- The typical conventional facilities are assumed to have modern, high performance HVAC systems. Most actual facilities are unlikely to have such economical existing systems. The profitability of the incremental investment in cogeneration when installation is combined with the simultaneous upgrading of the HVAC system will generally be slightly better than that indicated in Table 4-3.
- The cost of absorption chillers and associated cooling towers is included in the capital cost of the cogeneration system. Where facilities already have adequate absorption chiller and cooling tower capacity, payback periods will be reduced.
- Under current New York regulations, the use of interruptible gas by large customers requires dual fuel capability and some fuel storage capacity. Here, capital costs were not increased to account for this requirement. It is possible that this can add 10% or more to the total installed costs of the cogeneration system assumed in this study, depending on the facility's existing fuel storage capability and the planned operating strategy

**Table 4-3. Cogeneration Economics for Typical Facilities**
*Units: Capital costs in 1982 $. First year savings in 1985 $.*

| | *Residential* | *Commercial* | *Hospital* | *University* | *Hotel* |
|---|---|---|---|---|---|
| Peak Electric Demand (mw) | 3.2 | 5.7 | 2.5 | 3.6 | 2.6 |
| Cogeneration Capacity (mw) | 2.0 | 4.0 | 2.0 | 2.0 | 2.0 |
| Capital Cost ($1000's) | 2200 | 3900 | 1800 | 2200 | 2200 |
| Pre-tax First Year Savings ($1000's) | 357 | 679 | 404 | 695 | 356 |
| Pre-tax Payback (yrs) | 6.6 | 5.8 | 5.0 | 3.4 | 6.0 |
| Pre-tax IRR (%) | 13.5 | 18.5 | 16.0 | 28.5 | 15.5 |
| After-tax First Year Savings ($1000's) | 565 | 1023 | NA | NA | 529 |
| After-tax Payback (yrs) | 5.5 | 5.0 | NA | NA | 5.0 |
| After-tax IRR (%) | 8.1 | 12.3 | NA | NA | 9.6 |

*Notes:*
1. Pre-tax calculations include property tax for residential, commercial, and hotel and exclude it for hospital and university.
2. After-tax calculations assume owner is in highest tax brackets.

for periods of interruption. These additional costs increase payback periods over those shown.

- While it was assumed that cogeneration systems would use reciprocating engines, alternative prime movers may be appropriate in some cases, especially for larger installations. In particular, gas turbines, steam turbines, low speed diesels burning residual oil, and dual gas turbine/steam turbine systems may be more economical in large facilities.
- The investments in the residential, commercial, and hotel cogeneration systems are assumed to be made by private corporations in the highest income tax brackets. Also, these facilities are assumed to pay full property taxes on the cogeneration equipment (which in New York City are over 5%). An investor in a lower tax bracket would see a longer after-tax payback period. Lowering the property tax rate shortens the payback period.
- A backup generator equal in capacity to a single cogenerating engine-generator is assumed to be required at an installed cost of $400/kw for all of the facilities except the hospital (which was assumed to have back-up capacity in place). For nonhospital facilities already having such backup generators, the total capital cost of converting to cogeneration will be correspondingly less, reducing the payback period.

Because of these kinds of variations in facility characteristics and the tax status of potential investors, the investment economics in specific situations will differ from those shown in Table 4-3. While the majority of New York building cogeneration opportunities in the 1–3 mw size category are unlikely to prove attractive to investors, it is reasonable to expect that in some cases payback periods will fall in a more acceptable range.

### *Sensitivity of Return to Various Parameters*

The sensitivity of the IRR to variations in capital costs, fuel types, fuel escalation rate, electricity costs, and whether or not a property tax is imposed, was examined for the commercial and the hospital facilities (see Figure 4-2). Capital costs were varied 20%

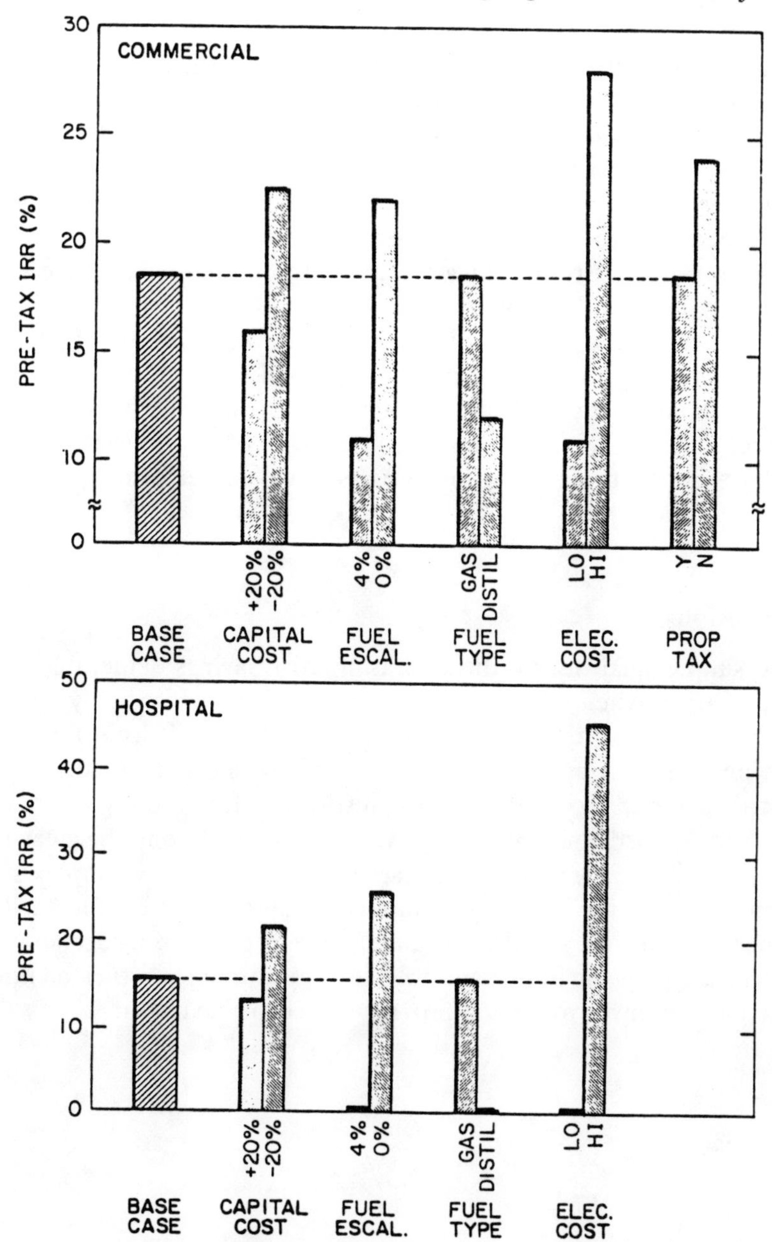


**Figure 4-2. Effect of Changes in Various Parameters on Internal Rate of Return for Cogeneration Investment in Commercial and Hospital Facilities**

above and below the base case level. Distillate fuel, with a 1982 cost of $7 million Btu, was compared with natural gas at $5.70 million Btu, the base case assumption. Fuel escalation rates of 4% and 0% were compared to the base case assumptions of 2%. Alternative electricity price projections were tested incorporating different assumptions regarding the utility's generating mix and fuel costs. The results of the sensitivity study indicate that all these factors can have a significant effect on the economics.

The capital cost variations incorporated into the sensitivity analysis are a good reflection of the range in costs likely to occur as a consequence of different facility and cogeneration systems characteristics. Hence, it is clear that some facilities will have somewhat better IRR's than do the average facilities assumed in performing the typical facility analysis.

## Conclusions

A simple analysis of the operating cost savings achievable with cogeneration when electricity rates are 12–15¢/kwh is likely to lead to the conclusion that large buildings in New York City offer excellent investment opportunities. However, the more detailed analysis described here, accounting in a realistic way for standby and interconnection costs and for the dynamic interplay among the electrical load, thermal load, and electric tariff, suggests a more complex situation. Site-specific load and equipment characteristics, the facility owner's tax status, and the size and operating mode of a cogeneration system all play an important role in determining whether or not a cogeneration investment will provide an acceptible return.

# Chapter 5

# Planning Cogeneration Systems for Hospitals

***James R. Clements***

In September 1982, United Enertec, Inc. and Martin Cogeneration Systems were contracted by the Gas Research Institute (GRI) for Phase I of the Research and Development of a Packaged Cogeneration System for Hospitals. This project was initiated as part of GRI's near-term effort to develop packaged cogeneration for early market entry while working on long-term component improvement. The objective of Phase I was to design and evaluate a pre-engineered, packaged gas-fired cogeneration system for mid-sized hospital facilities and to assess the potential thereof. This chapter describes the activities accomplished in the GRI project, including a description of the hospital energy market and the benefits of cogeneration in hospitals.

## The Hospital Market

Providing high-quality medical care requires enormous amounts of energy. Hospitals use nearly twice the energy consumed in the average American home for each hospital bed. Health care institutions collectively consume nearly 15% of the energy used in all commercial buildings in the U.S.

Hospitals today are viewing energy conservation as an important part of a total health-care cost containment program. There are, however, several unique factors that complicate energy-reducing efforts in health care institutions. The major factors include:

- Stringent enforcement of construction codes
- Scope and sophistication of hospital energy-using systems and equipment
- Limitation on capital.

*Stringent Enforcement of Construction Codes*

Current state and federal standards require hospitals to maintain strict tolerances in temperature and humidity levels in order to ensure optimal patient care. Federal regulations also require that facilities be divided into positive and negative air pressure zones to reduce the possibility of cross-contamination. Although some modifications in the current standards are being proposed, there is a general lack of authoritative technical information to support or refute the existing codes. As a result, they continue to require institutions to exhaust considerable quantities of conditioned air, and replace it with outside air that must be heated or cooled.

*Scope and Sophistication of Energy-Using Systems*

Because hospitals are sophisticated energy users, a "technical mystique" has developed around the use and reduction of energy within the facility. Hospitals offer varying scopes and intensities of patient care services, which affect energy use in different ways in different institutions. Large quantities of energy are needed for the support services—that are essential to the efficient functioning of a hospital.

Energy is needed to power the diagnostic, therapeutic, and monitoring equipment that is vital to effective delivery of modern medical services. The typical energy conservation measures that are applied to many types of buildings do not always represent significant energy-saving options for hospitals. There is a limit to how low thermostats can be set for sick and elderly patients without adversely affecting their well-being. Air conditioning is not a luxury in patient care areas, since it has definite therapeutic value. Obviously, maintaining tight control over energy use in hospitals is a demanding engineering responsibility.

*Limitations on Capital*

At present, most hospital revenues come from third-party payers —nonprofit reimbursement or prepayment plans (Blue Cross Plans), government programs (primarily Medicare and Medicaid), and commercial insurance companies. Most capital for investment in hospital plant facilities and equipment is obtained from traditional commercial markets (mortgages, taxable bonds, municipal bonds, philanthropic contributions and public programs). To some extent, this financial environment limits the hospital's ability to achieve greater energy efficiency. Most hospital revenues still flow from third-party payers that reimburse on the basis of retrospective costs. In these cases, facilities do not reap the full benefits of their energy conservation investments.

Reducing energy consumption in hospitals is further complicated because few of the existing facilities were designed with energy conservation in mind. Some 90 percent of the nation's health care institutions were built prior to the 1973-74 oil embargo, when energy was seemingly abundant and inexpensive. Each hospital's energy consumption pattern and corresponding energy budget will be unique. The graphics presented in this section are intended to provide insight into the distribution of energy costs within the typical hospital, and prove the viability of packaged cogeneration systems for hospitals.

**Hospital Energy Profile**

Figure 5-1 illustrates in the form of a pie chart the annual energy *budget* of a typical hospital by various departments and types of equipment. The cost breakdown applies to hospitals with either electric or absorption chillers. The largest portion of the yearly energy cost, 40-60%, is expended for the operation of heating, ventilating and air conditioning (HVAC) equipment, such as boilers, chillers, fans and pumps. This category also presents the greatest potential for energy savings.

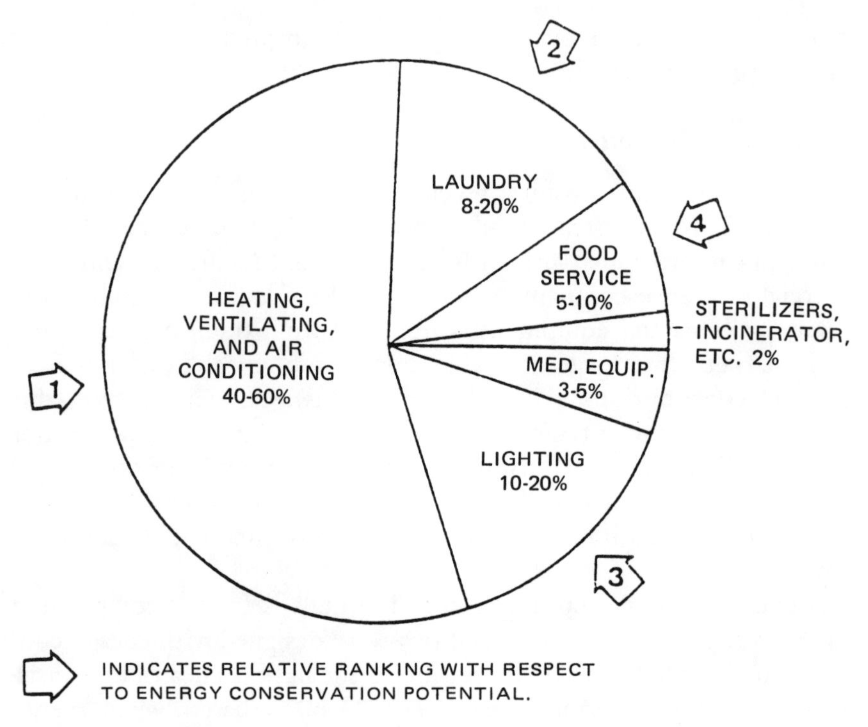


**Figure 5-1. Typical Breakdown of Hospital Energy Costs**

Figures 5-2 and 5-3 illustrate the typical hospital energy distribution load profile. These have been derived from data obtained from 350 facilities owned and operated by the Hospital Corporation of America, 80 Veterans Administration facilities in the states selected for this study, and 165 hospitals in the Georgia Power Company service territory, plus information obtained from the American Hospital Association.

Electrical and fuel costs now constitute 8 to 15% of the hospital's total operating budget. The hospital power plant designs have historically been designed based on first cost only with disregard for life-cycle cost or system efficiency.

The typical daily electrical load profiles as shown in Figure 5-4 define the normal energy consumption patterns for a typical hot summer weekday, typical cold winter weekday, typical Sunday and

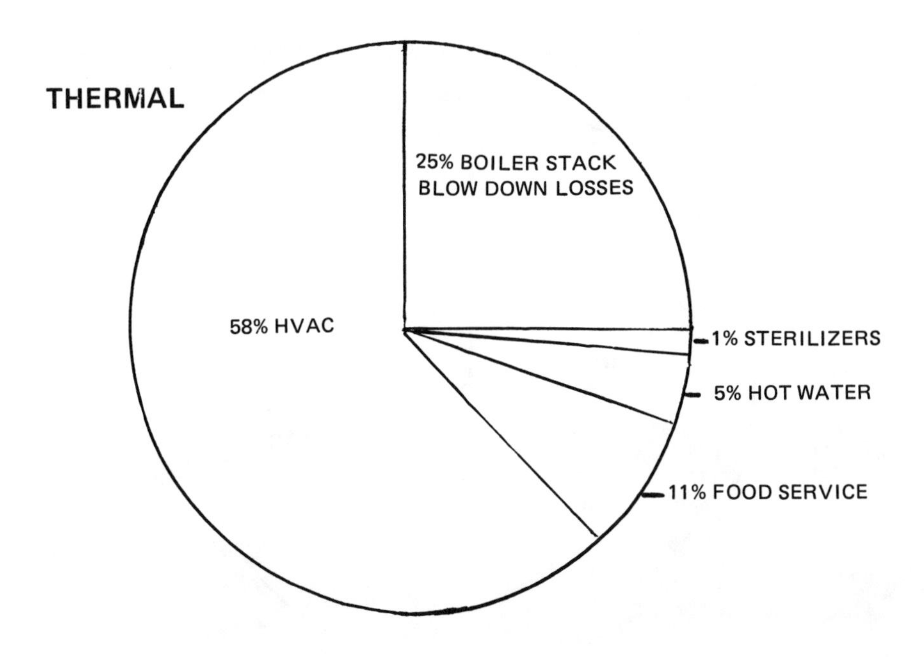


**Figure 5-2. Typical Thermal Load Distribution**

typical emergency engine-generator size to meet the hospital standby emergency codes NEC #700-5(b).

## Analysis of the Hospital Market for Cogeneration

The entire hospital market was surveyed through data provided by the American Hospital Association to determine that over 4,000 of the 6,965 U.S. hospitals are potential candidates for a single-unit package installation. Average hospital energy loads were determined on a thermal, electrical and cost basis.

Specific energy use data were obtained from over 400 hospitals, primarily those operated by Hospital Corporation of America (HCA) and the Veterans Administration (VA). On-site audits of 15 HCA facilities were conducted and daily, weekly, monthly and yearly energy profiles were developed.

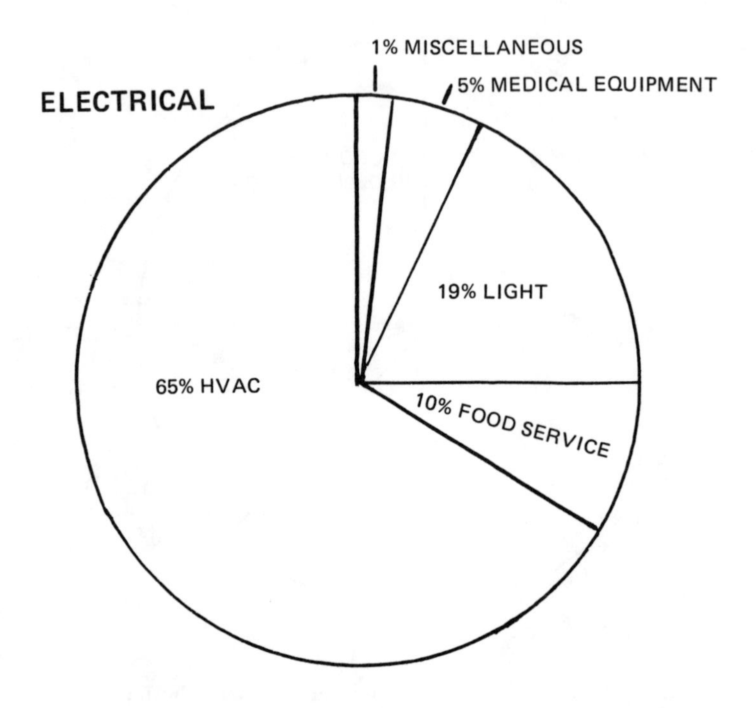


**Figure 5-3. Typical Electrical Load Distribution**

Excellent cooperation was received from HCA, which presently owns and operates over 350 "for profit" hospitals world-wide and has 100 additional facilities under some stage of construction.

In the on-site audits, siting and operational constraints were evaluated. Although certain facilities will have more favorable characteristics in these areas, no unresolved obstacles were identified.

Many of the facilities surveyed utilized four-pipe heating and cooling systems which are ideal for a cogeneration package. In this situation, the hot water Btus from the system can be directly injected into the hot water line, practically eliminating the need for a boiler to produce domestic hot water and space heating. The cold water output from the absorption air conditioning can be injected into the cold water line. In most cases, unused capacity in the existing cooling tower can be used for the absorption unit.

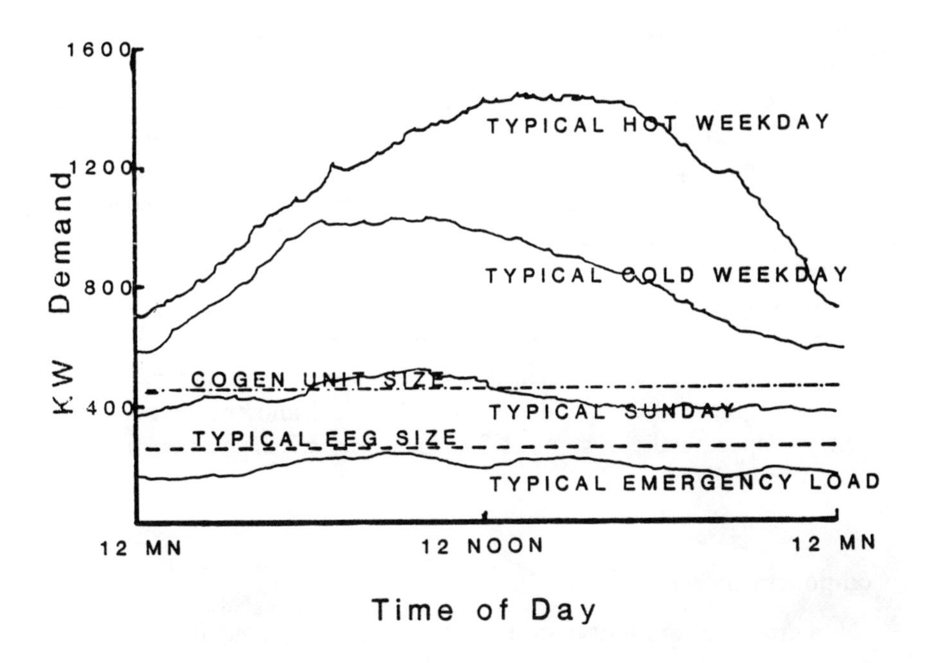


**Figure 5-4. Typical Electrical Load Profile**

All of the facilities housed existing emergency generators of 300+ kw capacity. These units would be useful to avoid demand charges when the cogeneration unit is down for maintenance. Each of the facilities had space for the package in the general proximity of the existing power room. Especially noise-sensitive areas did not appear to present a problem for the designed package noise level. Each facility appeared eager to do something to reduce its spiraling energy costs.

## Module Design and Performance Analysis

A general design of 300–450 kw capacity was identified to provide the widest application for the hospital market. Design and analytical studies were accomplished to determine optimal cost/performance characteristics for each component and the entire system. A detailed design was completed and alternative component manufacturers were identified and contacted.

An isochronous generator was selected to parallel with the electrical utility in order to provide additional emergency backup, as well as provide power factor control. Heat recovery in the form of 210° hot water was chosen as the basic thermal medium; however, a steam separator can be easily added to provide 40 psi steam if required.

A housing and base 38 ft long, 8 ft wide and 10 ft high was chosen to minimize cost and area required. The control system was designed to provide all pre-alarm, alarm and automatic shutdown capabilities. In addition, it was designed to sense electrical and thermal requirements to operate the system at full load only when the thermal output can be utilized.

**Component Selection**

Numerous component manufacturers were contacted and evaluated on cost, reliability and service capability. Excellent cooperation and interest was received from the manufacturers. Special "one-time only" pricing for the prototype, plus offers to conduct research to improve component performance, were received from each vendor.

The entire system was divided into the following subsystems:

- Prime Mover
- Generator
- Switchgear
- Control System
- Heat Recovery
- Housing and Base
- Absorption Air Conditioning

Component selections were based on current cost and current technology; however, further technical improvement and cost reduction opportunities were identified.

**Commercial Feasibility Assessment**

A costed bill of materials was prepared based on current OEM costs, predicted mass-production costs and predicted improved technology costs. The range of these costs varied by over 25%. Maintenance costs and other operating costs were identified. The operating cost characteristics were integrated with the system's anticipated performance data to determine the financial results of installing a system in five separate markets. The results varied widely due to specific utility rate structures with predicted annual savings ranging between $103,000 and $271,000. An analysis through 1995 was made using predicted escalation of fuel and electricity rates. All markets indicated adequate return on investment, with paybacks ranging from 1.5 to 5 years.

Benefits of the system to the utility ratepayer, gas industry and the hospitals were identified. As the 450 kw system consumes 49,056 MCF of gas at full load, large commercialization could lead to the consumption of 198 BCF per year. In addition, the normally weak demand for gas during the summer months would predictably be the highest usage of the system.

A summary of the system's anticipated financial performance follows. The installed Capital Cost of 450 kw package is:

| *Present Cost* | *Mass Production Cost* | *Improved Technology Cost* |
|---|---|---|
| $ 464,490 | $402,350 | $345,150 |
| $1,032/kw | $894/kw | $767/kw |

In order for the purchaser of a Hospital Cogeneration Package to pay back this substantial investment, the load profile and the utility rate characteristics must meet certain criteria. The hospital and energy load demand must utilize the full thermal and electrical output of the package for at least 6,000 full load hours. In addition, the differential between the price of the fuel purchased and the electricity displaced must be equivalent to $.05 per kwh ($15 per million Btu) generated.

Thus, the site selected for the prototype will be in an area that has gas priced at no greater than $5.00/MCF and electricity at no less than $.08/kwh. These prices are quite common in California and New York, and they are increasingly occurring throughout the rest of the U.S.

Under a scenario of $5.00 gas, $0.08/kwh electricity, and offset thermal requirements at the $5.00 rate and 75% efficiency, the following economic performance will be anticipated for the testing period:

| | | |
|---|---|---|
| Electricity Cost Saved<br>(8,000 hrs) (425 net kw) ($0.08/kwh) | = | $272,000 |
| Thermal Energy Saved (assuming 100% utilization)<br>(8,000 hrs) (2.972 MMBtu/hr)<br>($5.00/MMBtu) (0.75) | = | $158,506 |
| Predicted Savings | = | $188,506 |

In many instances, the hospital's thermal requirements are not continuously sufficient to utilize all of the system's thermal output. Therefore, in order to utilize the thermal energy year-round, 72 tons of absorption air conditioning have been added to the package for the equivalent of 5,000 full-load hours per year.

During the warmer months, it is anticipated that only half of the thermal output can be utilized for domestic hot water and limited space heating, and therefore absorption air conditioning utilizing 46% of the output has been included. This tonnage can be reduced or increased depending on the site-specific requirement. The cost or savings of this charge would be approximately $900/installed ton.

With the absorption air conditioning added, the financial results become:

| | | |
|---|---|---|
| Electricity Generated Savings<br>(8,000 hrs) (425 kw) ($0.08/kwh) | = | $272,000 |
| A/C Electricity Savings<br>(5,000 hrs) (72 tons) (1.25 kw/T)<br>(0.08/kwh) | = | $36,000 |
| A/C Thermal Energy Utilized<br>(5,000 hrs) (72 T) (19,000 Btu/T) | = | 6,840 MMBtu |

Thermal Savings

$$\frac{(8{,}000 \text{ hrs})\ (2.972 \text{ MMBtu/hr})\ (6{,}840 \text{ MMBtu})}{.75} = \$112{,}907$$

Cost of Gas

(8,000 hrs) (5.60 MMBtu/hr)
(5.00/MMBtu) =($224,000)

Maintenance

(.005/kwh) (8,000 hrs) (450 gross kw) =($ 18,000)

Predicted Savings = $178,907

Thus, the net savings will be decreased by $9,599; and, the capital requirement will be increased by approximately $50,000. With the installed target of $900/kw, the cost of $405,000 (450 kw at $900/kw), would be paid back in 2.28 years.

In conclusion, the system's financial performance has been calculated at optimal system performance with favorable utility rates. However, if this performance can be attained or improved upon, and if the improved system cost can be accomplished, the Hospital Cogeneration Package should have outstanding market potential throughout the country.

**Continuing Activities**

The results of the Phase I effort described here show that a cost-effective cogeneration package can be developed for a significant number of hospital facilities. The continuing activities on this project will be to confirm the assumptions that lead to the Phase I conclusion, as well as to point out specific research that may lead to improved system cost and performance. Several of the key variables to be confirmed in Phase II are:

- First Cost of Hardware
- Installation Cost
- Maintenance Cost
- Reliability
- System Efficiency
- Actual Savings

Each of the above must meet specific goals for the system to be commercially viable. To determine whether this is possible, a prototype system will be built and each factor tested. The prototype will then be installed in an actual hospital and operated for at least one year.

Phase II is expected to take approximately 24 months from inception to installation and monitoring. The following steps will be accomplished:

1. Monitoring and selection of a hospital before installation in order to establish precogeneration operating characteristics.
2. Developing detailed production drawings and building the prototype.
3. Testing the package in a controlled environment at the factory.
4. Installing the prototype at the hospital.
5. Monitoring and evaluating specified variables and operating results.

As the data are gathered, compiled and analyzed, a commercialization plan will be developed. The Caterpillar dealer network will be utilized for marketing, installing and servicing the system.

The successful development of the Hospital Cogeneration Package would produce significant sales of equipment and natural gas. Phase I indicated the potential for as many as 5,000 hospital installations. Each unit would consume 44,800 MCF of natural gas per year, less a potential maximum reduction of 26,298 MCF by thermal recovery, leading to a net increased consumption of 18,502 MCF/year. Thus, the increased gas consumption would total 83,259,000 MCF/year.

However this increased gas consumption is understated by the amount of thermal energy that offsets electricity (such as air conditioning) or fuel oil (certain boiler applications). It is also anticipated that modifications to the Hospital Cogeneration Package would have significant application for nursing homes, food processing plants, and numerous other operations requiring hot water or low pressure steam. These markets would open up thousands of additional opportunities for increased equipment and natural gas sales.

# Chapter 6

# Computerized System Designs

**Fred H. Kindl**

This chapter is not going to describe how to feed in a few design parameters, push a button, and by the magic of the computer produce a set of installation drawings for a cogeneration system. There may indeed be a market for such a system, but cogeneration is not quite that cut and dried.

This chapter will discuss how the computer can be used to:

- determine whether cogeneration has a chance under a range of circumstances at a given site, and to
- identify the optimum conceptual design for a cogeneration system given the site specific limitations.

## The Objective

The broad goal is to ascertain if it is possible to save money by cogeneration. To be more specific, there is a site-by-site economic question of determining the rate of return the savings will bring if money is invested in a cogeneration system.

### *Information Requirements*

To meet this objective for a given site two numbers are needed: first, the size of the investment, and second, a prediction of the savings. Furthermore, the second number is not independent of the first. A different system will require a different capital investment and yield a different savings. With a given system in mind it is usually far easier to come up with a reasonable value for the investment than it is for the savings. The investment is the straightforward sum of the required equipment plus installation.

The savings, on the other hand, are the result of a complex interaction of power and heat requirements, fuel and power costs, the kind of equipment purchased, and how we operate it; and all of these are in the future with the added uncertainty that entails. Savings, of course, are also affected by interest charges, taxes, investment credits, depreciation allowances, etc., but these are reasonably simple to quantify once the investment is defined and the annual savings determined.

The basic force that drives any energy system modification is the predicted annual savings in energy costs resulting from a change. Our objective then becomes to determine these as accurately as possible and, recognizing they are a difference between the "before" and "after" operating costs (usually larger numbers than their difference), it is imperative that the before and after expenses are calculated with considerable precision. Is it really any wonder that in the absence of such knowledge many corporate decision-makers would rather not gamble on cogeneration at their plants?

If cogeneration is to be believed on a broad scale, it is necessary to have a means for predicting with reasonable certainty what the savings will be from changing from the traditional modes of separate heat and power generation to cogeneration. Computer simulation is the means to develop the needed before and after estimates of operating costs that predict the flavor of the cogeneration pudding.

## Major Factor Affecting Operation

There are four principal factors that affect the cost of operation of an energy supply system:

- Plant loads
- Plant equipment
- Operating mode
- Economic environment

These factors affect the cost of operation both by themselves and in synergistic combination.

Let's look at these factors in more depth.

There are three kinds of *plant loads:*

- Electrical
- Thermal (with possible multiple temperature levels)
- Mechanical

and they vary:

- Hourly
- Daily
- Seasonally
- Independently among the types.

It is essential that loads be modeled in detail to accurately determine the impact of a cogeneration installation. Figure 6-1 is a typical set of hourly load curves used to model one month of the year. When the modeling is completed, a summary of each month's characteristics is generated as shown in Table 6-1 and an annual load duration curve is produced as shown in Figure 6-2.

Note that while the load duration curve gives some information about equipment requirements, it is inadequate for evaluating cogeneration applications which require information about the coincidence

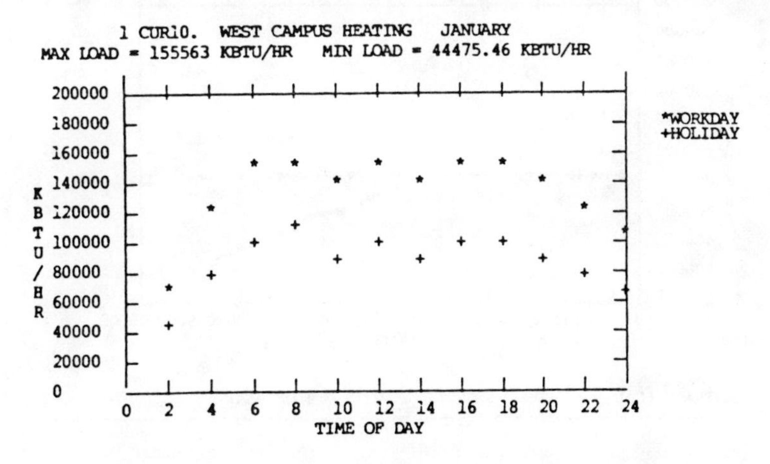


Figure 6-1. Typical Hourly Loads

**Table 6-1. Summary of Load Information**

CUSTOMER = SAMPLE ALTERNATIVE = EXISTING CONDITION
LOAD = H1/WEST CAMPUS HEATING/ (FROM FILE L0010)
FOR YEAR: 1981 PERIODS: ALL

SUMMARY LOAD INFORMATION

| *Period* | *Avg. Daily Use (KBTU)* | *Avg. Hourly Load (KBTU/HR)* | *Total Hours* | *Total Use (KBTU)* |
|---|---|---|---|---|
| 1 Jan | 2894215.49 | 120592.31 | 744 (31 Days) | 89720680.24 |
| 2 Feb | 3348696.92 | 139529.04 | 672 (28 Days) | 93763513.7 |
| 3 Mar | 2651996.4 | 110499.85 | 744 (31 Days) | 82211888.33 |
| 4 April | 1714080.02 | 71420 | 720 (30 Days) | 51422400.61 |
| 5 May | 807938.69 | 33664.11 | 744 (31 Days) | 25046099.49 |
| 6 June | 1535681.37 | 63986.72 | 720 (30 Days) | 46070440.97 |
| 7 July | 1670474.67 | 69603.11 | 744 (31 Days) | 51784714.89 |
| 8 Aug | 1872333.8 | 78013.91 | 744 (31 Days) | 58042347.9 |
| 9 Sept | 1818383.09 | 75765.96 | 720 (30 Days) | 54551492.68 |
| 10 Oct | 1396672.07 | 58194.67 | 744 (31 Days) | 43296834.09 |
| 11 Nov | 2093206.08 | 87216.92 | 720 (30 Days) | 62796182.4 |
| 12 Dec | 2629601.62 | 109566.73 | 744 (31 Days) | 81517650.21 |
| TOTAL | 2028011.63 | 84500.48 | 8760 (365 Days) | 740224245.51 |

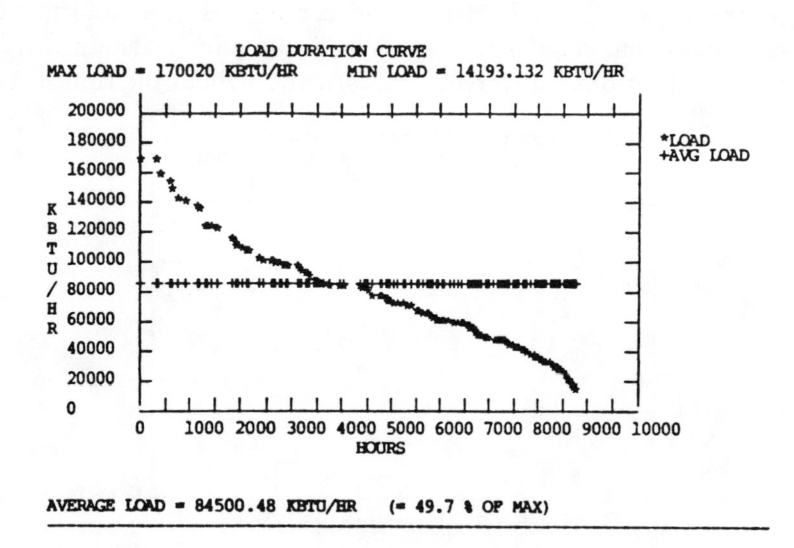


**Figure 6-2. Annual Load Duration Curve**

in time between required heating and power levels. Load modeling with hourly heat and power curves is necessary for accurately evaluating cogeneration applications.

*Plant Equipment* characteristics must be defined and accounted for in the analysis recognizing that:

- *Capability* may vary with ambient pressure and temperature
- *Efficiency* will vary with load and ambient conditions
- *Heat recovery potential* is affected by site conditions
- *Availability and reliability* is often less in practice than claimed by equipment manufacturers.

*Operating Modes* may have a favorable or unfavorable impact on actual operating savings. While the most economical mode would normally be chosen, it is important to determine operating costs on the basis of the one which will be used, particularly if local considerations dictate something other than the most economical approach. Some typical operating modes are:

- Thermal tracking
- Electrical tracking
- Isolated
- Peak shaving
- Economically optimized
- Energy supply reliability maximized.

The *Economic Environment,* of course, must be included in the analysis to be certain the technical alternative that has been identified as sound is really cost effective. This includes such items as:

- Fuel cost
- Fuel availability
- Power cost
- Excess power value
- Cost of money
- Future value of all of the above.

Having properly identified and accounted for the necessary inputs, Encotech uses the power of the digital computer to model the system under investigation, dispatch the available equipment to meet the hourly load demands, and calculate the cost of fuel, maintenance, operators, fixed charges, and purchased and sold power to meet the energy requirements. The summation of these charges for a year's time is the annual cost of operations.

However, one year is not enough. Future changes in load requirements or energy costs will have a major impact on the desirability of constructing an alternative energy supply system. Therefore, the energy supply costs and load requirements are projected into the future, and a cost of operations is calculated for each year of the study. A typical result is shown in Table 6-2.

The "bottom line" is: will this alternative arrangement give a satisfactory return on the necessary investment? Following completion of the cost of operation analysis, a financial program determines the after-tax net cash flow for each year of the study using present worth principles, and calculates a return on the investment sometimes called the Internal Rate of Return, or just ROI. A typical financial analysis is shown in Table 6-3.

It should be stressed that as mentioned earlier, these figures are differences between the cogeneration alternative under consideration and the existing plant base case. Sensitivity analyses then are used to highlight how susceptible the economic results are to changing conditions.

Experience has demonstrated that cogeneration is thermodynamically sound and technically practical. In many cases it is also economically correct and a means to obtain substantial reductions in annual energy bills. Whether this latter is true for a particular installation can only be determined by a sophisticated analysis.

So far, the analysis in this chapter has been presented as though there were only one cogeneration alternative under consideration. Obviously, that is not the case. In general, several different cogeneration systems are evaluated, and for this engineering judgment and experience come into play.

By a process of examining various alternative cogeneration schemes and looking at the resulting sensitivity analyses, a view begins to develop of which factors are most important for a given set of circumstances and which prime mover is best for a given site. This is not an automatic computerized, linear programming optimization,

**Table 6-2. Example of Cash Flows**

SUMMARY CASH OUTFLOWS AND (INFLOWS) IN THOUSANDS OF DOLLARS
(CASE = 11 DATE = 821212 174801724)

| | Year | *Fuel Cost* | *VOM* | *FOM* | *Fixed Charges* | *Electricity Bought* | *Electricity Sold* | *Subtotal* | *Equipment* |
|---|---|---|---|---|---|---|---|---|---|
| 1 | 1981 | 10976.95 | 446.38 | 433 | 13.62 | 13160.81 | (0) | 25030.76 | 0 |
| 2 | 1982 | 24105.35 | 260.03 | 641.3 | 314.4 | 4595.95 | ( 984.93) | 28932.11 | 14971 |
| 3 | 1983 | 27723.57 | 275.63 | 679.78 | 333.27 | 5451.44 | (1168.27) | 33295.41 | 0 |
| 4 | 1984 | 31884.88 | 292.17 | 720.56 | 353.26 | 6466.18 | (1385.73) | 38331.32 | 0 |
| 5 | 1985 | 36670.8 | 309.7 | 763.8 | 374.46 | 7669.79 | (1643.67) | 44144.87 | 0 |
| 6 | 1986 | 42175.08 | 328.28 | 809.63 | 396.93 | 9097.44 | (1949.62) | 50857.74 | 0 |
| 7 | 1987 | 48505.56 | 347.98 | 858.2 | 420.74 | 10790.84 | (2312.52) | 58610.81 | 0 |
| 8 | 1988 | 55786.25 | 368.86 | 909.7 | 445.99 | 12799.45 | (2742.97) | 67567.27 | 0 |
| 9 | 1989 | 64159.76 | 390.99 | 964.28 | 472.74 | 15181.94 | (3253.55) | 77916.16 | 0 |
| 10 | 1990 | 73790.14 | 414.45 | 1022.13 | 501.11 | 18007.91 | (3859.17) | 89876.58 | 0 |
| 11 | 1991 | 84866.04 | 439.32 | 1083.46 | 531.18 | 21359.9 | (4577.51) | 103702.39 | 0 |
| 12 | 1992 | 97604.44 | 465.68 | 1148.47 | 563.05 | 25335.83 | (5429.57) | 119687.89 | 0 |
| 13 | 1993 | 112254.86 | 493.62 | 1217.38 | 596.83 | 30051.84 | (6440.23) | 138174.29 | 0 |
| 14 | 1994 | 129104.32 | 523.23 | 1290.42 | 632.64 | 35645.69 | (7639.02) | 159557.28 | 0 |
| 15 | 1995 | 148482.87 | 554.63 | 1367.85 | 670.6 | 42280.78 | (9060.94) | 184295.78 | 0 |

## Table 6-3. Typical Results of Financial Analysis

RESULTS OF FINANCIAL ANALYSIS
(FOLLOWING VALUES ARE DIFFERENTIALS IN 1000 DOLLARS
OF CASE 21 VERSUS CASE 19.)

| *Year* | *Equipment* | *Debt* | *Depreciation* | *Federal Taxes* | *State Taxes* | *Local Taxes* |
|---|---|---|---|---|---|---|
| 1 | 14971 | 9981 | 2994 | 286 | 69 | 21 |
| 2 | 0 | 0 | 2395 | 845 | 204 | 63 |
| 3 | 0 | 0 | 1916 | 1404 | 339 | 105 |
| 4 | 0 | 0 | 1533 | 1979 | 478 | 148 |
| 5 | 0 | 0 | 1226 | 2587 | 625 | 193 |
| 6 | 0 | 0 | 981 | 3244 | 784 | 242 |
| 7 | 0 | 0 | 785 | 3965 | 958 | 296 |
| 8 | 0 | 0 | 628 | 4767 | 1151 | 356 |
| 9 | 0 | 0 | 502 | 5667 | 1369 | 423 |
| 10 | 0 | 0 | 402 | 6685 | 1615 | 499 |
| 11 | 0 | 0 | 0 | 7971 | 1925 | 595 |
| 12 | 0 | 0 | 0 | 9264 | 2238 | 692 |
| 13 | 0 | 0 | 0 | 10752 | 2597 | 803 |
| 14 | 0 | 0 | 0 | 12463 | 3010 | 931 |
| 15 | 0 | 0 | 0 | 14430 | 3486 | 1078 |

| *Year* | *Operating Savings* | *Total Taxes* | *Loan Repayment* | *Interest* | *Investment Tax Credit* | *Net Cash Flow* | *Equity* |
|---|---|---|---|---|---|---|---|
| 1 | 4927 | 377 | 245 | 1220 | 1497 | 4582 | 4990 |
| 2 | 5687 | 1112 | 277 | 1188 | 0 | 3110 | 0 |
| 3 | 6563 | 1847 | 314 | 1152 | 0 | 3250 | 0 |
| 4 | 7572 | 2605 | 355 | 1111 | 0 | 3502 | 0 |
| 5 | 8733 | 3405 | 401 | 1064 | 0 | 3863 | 0 |
| 6 | 10071 | 4270 | 454 | 1012 | 0 | 4336 | 0 |
| 7 | 11610 | 5219 | 513 | 952 | 0 | 4926 | 0 |
| 8 | 13383 | 6274 | 580 | 885 | 0 | 5644 | 0 |
| 9 | 15423 | 7459 | 656 | 809 | 0 | 6499 | 0 |
| 10 | 17772 | 8799 | 742 | 724 | 0 | 7508 | 0 |
| 11 | 20475 | 10492 | 839 | 627 | 0 | 8518 | 0 |
| 12 | 23586 | 12194 | 949 | 517 | 0 | 9927 | 0 |
| 13 | 27166 | 14152 | 1073 | 393 | 0 | 11549 | 0 |
| 14 | 31286 | 16404 | 1213 | 252 | 0 | 13417 | 0 |
| 15 | 36027 | 18994 | 1372 | 94 | 0 | 15568 | 0 |

SIMPLE PAYBACK FOR EQUIPMENT INSTALLED IN YEAR 1 IS: 3.04 YEARS.

RETURN ON INVESTMENT IS 79.375%.

but rather optimization in the sense of judgmentally homing in on the best choice by having examined the options in considerable depth.

This approach can be used to determine simply whether cogeneration has a chance at a given site, and also to finally selecting the optimum equipment for that location. This same analytical framework can be used for rapid screening studies.

**The Computer Model**

The computer model consists of:

- Two main calculation programs
- 18 data files
- Three auxiliary programs
- Numerous output summaries

as shown in Figure 6-3. Operationally, the modeling process follows a logic flow as indicated in Figure 6-4. The most important output of the "Cost" program is the tabulation of annual operating costs previously referred to and shown in Table 6-2. The financial program compares these data for two alternative arrangements and, making use of necessary financial information, calculates the savings from changing from one alternative to the other and the return on the necessary investment to make the change possible. A typical result was shown earlier in Table 6-3.

Full use of this program for an in-depth evaluation starts with detailed modeling of the electrical and thermal loads. Data on the hourly variations of thermal loads are frequently either incomplete, conflicting, or sometimes simply not available. At the beginning, the question is often whether cogeneration has any chance at all, and no one wants to go to the expense of a detailed study for what ultimately turns out to be an unpromising site.

There are many individual rules of thumb to reject the obvious misapplications of cogeneration. In order to go a step further and put some numbers on a potential site, some short cuts have been introduced to the program. As shown in Figure 6-3, a Curve Data Bank allows the program user to select load profiles typical of a given site

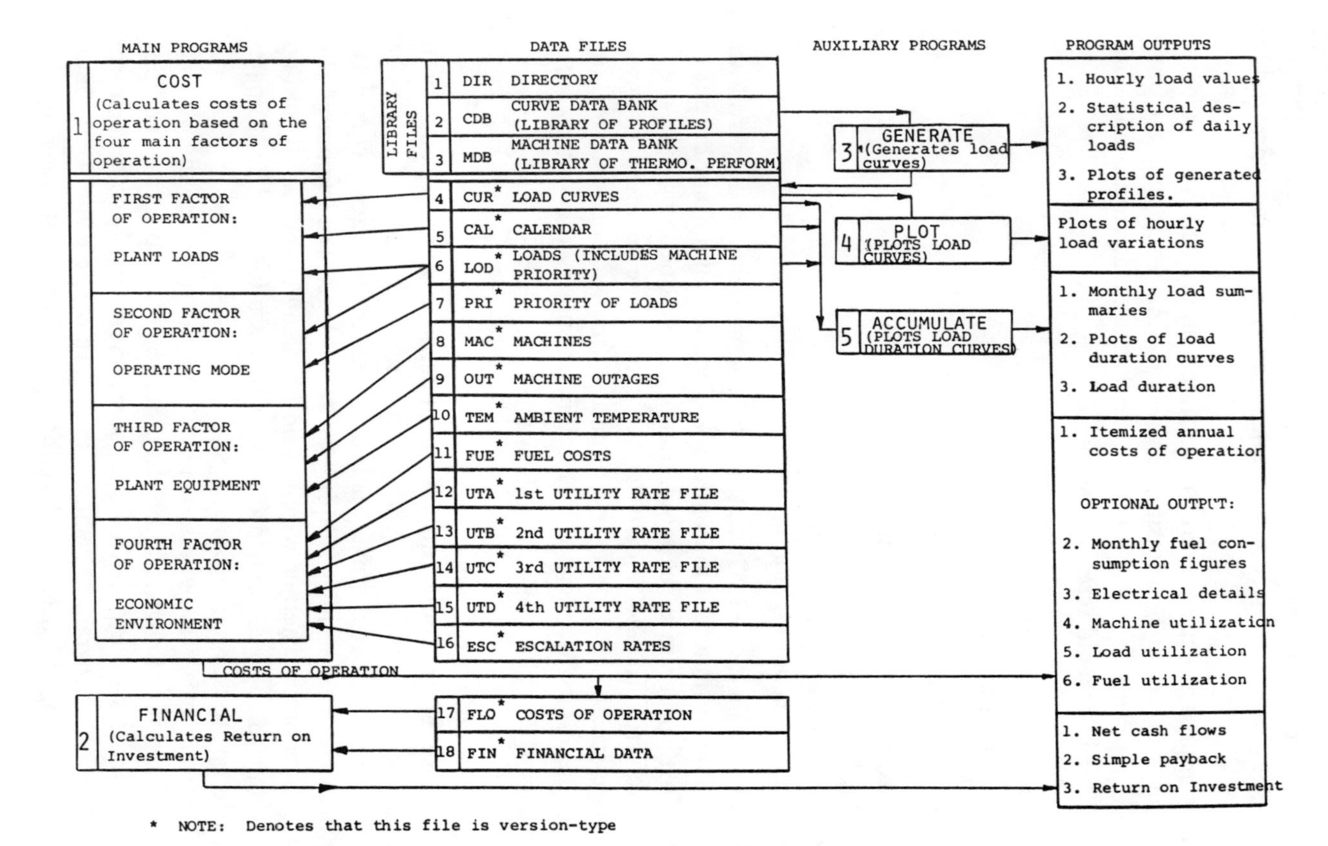


**Figure 6-3. Relationships Among the Five Programs, the 18 Data Files and the Outputs of the Energy Planning Program**

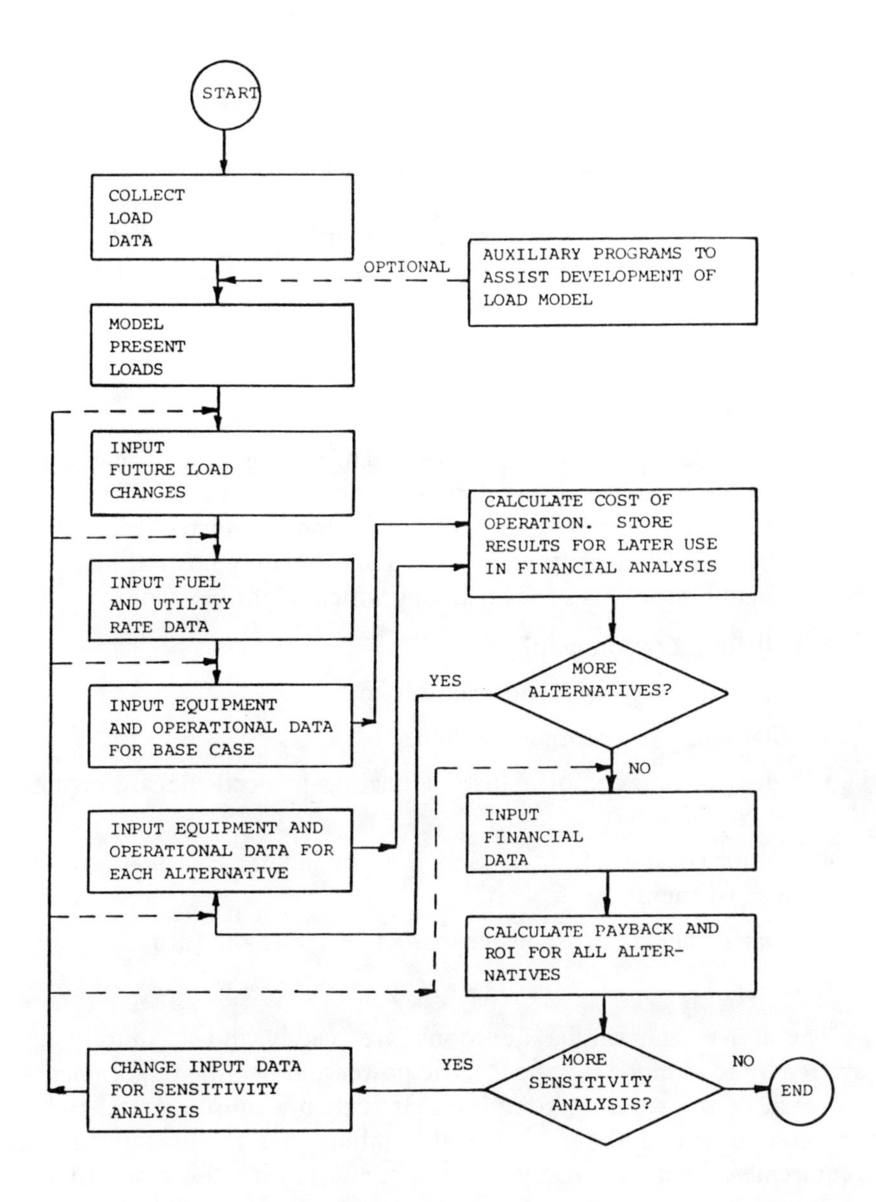


**Figure 6-4. Encotech's Cogeneration and Energy Planning Program Basic Flow Chart**

to eliminate the need for detailed load modeling, and a Machine Data Bank is available to describe the generic characteristics of many prime movers.

These features permit rapid first-cut screening evaluations of cogeneration feasibility. The American Gas Association is making this capability available to its member companies on a time share computer system to assist their gas marketing efforts, and it is available to other users as well.

## System Design

To conclude, consider a situation such as diagrammed in Figure 6-5. Here we have a major commercial installation in an urban area surrounded by several small commercial and industrial facilities. Given the task of identifying ways to reduce energy costs at installation A, many questions surface, among which might be:

- Will cogeneration help?
- Are heat and power loads suitably matched?
- What equipment should be used?
- Will development of a district heating project increase savings for complex A?
- Should complex B, C, D, and E all be included with A or only some of them?
- What if fuel costs rise faster than expected in the future?
- etc., etc.

The above, and similar questions, are readily and accurately answered by load modeling the electric power and heating requirements for each of the separate complexes. It is then a simple matter to let the computer add them in several combinations to meet the total requirements with a variety of equipment. Testing the more attractive combinations with a range of escalation rates for important costs is readily done to develop the sensitivity of the results to changing business conditions.

In summary the computer is a tool that does the extensive calculation work necessary to realistically model the anticipated system and approach the investment decision with confidence that the returns will make it worthwhile.

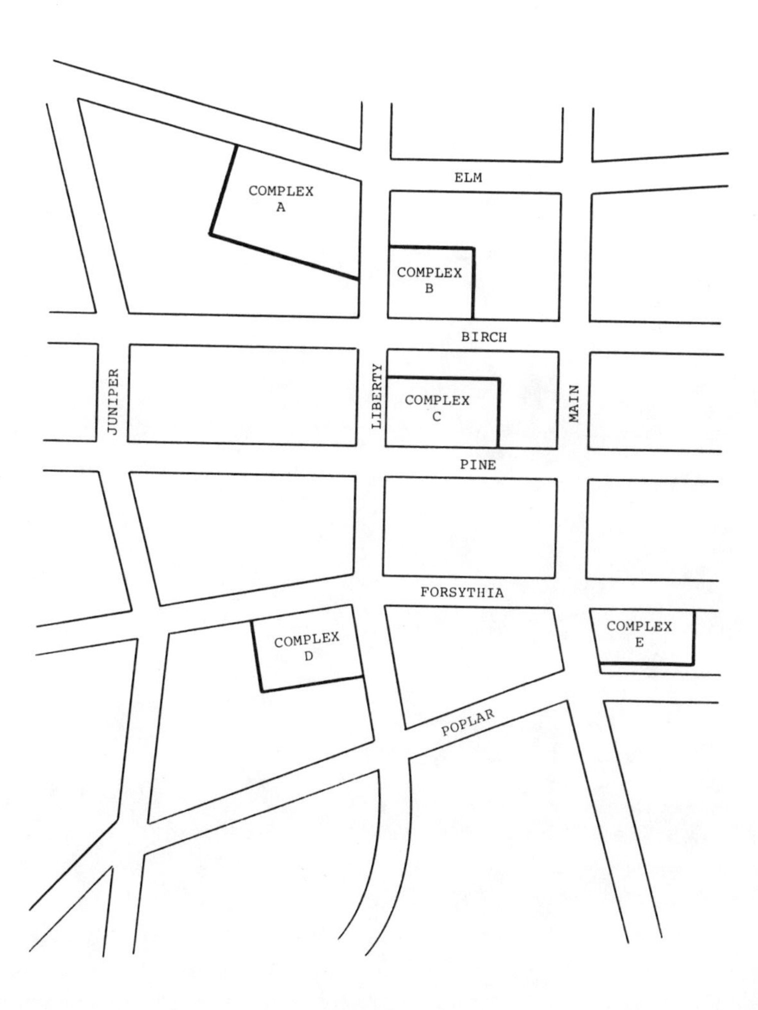


**Figure 6-5. Illustrative Site Arrangement**

# Chapter 7

# Evaluating Cogeneration Effectiveness

**M. P. Polsky and R. J. Hollmeier**

Recently, much has been written on the advantages of cogeneration (joint production of electrical and thermal energy). However, there has been little discussion on the methods of evaluation and the factors that affect fuel effectiveness. In this chapter several commonly used evaluation methods will be analyzed.

## Standard Evaluation Methods

The first method is based on the so-called fuel utilization factor (F.U.F.), or overall thermal efficiency ($\eta_t$), which is the ratio of all useful energy extracted from the cycle to the fuel input. The value of $\eta_t$ is defined by equation (7-1).

$$\eta_t = (N + Q_t)/Q_f \qquad (7\text{-}1)$$

This method is based on the first law of thermodynamics. Although useful for some illustrative purposes, this method (as shown below) cannot be used in most cases to evaluate the effectiveness of cogeneration. It is very important to note that greater fuel utilization does not necessarily mean a more efficient cycle. The majority of cogeneration power plants have an $\eta_t$ of 70 to 80%, whereas power plants supplying only thermal energy often have $\eta_t > 80\%$. Does this mean that the fuel utilization in the second case is more efficient?

Using $\eta_t$ for the comparison of different cogeneration plants is often totally unacceptable and may lead to incorrect results. For example, for plants with back pressure steam turbines, $\eta_t$ includes only mechanical and electrical generation losses as well as boiler leakage and other thermal losses. Steam turbine internal losses, which result in an increase in the exhaust enthalpy, are recovered in the form of thermal energy and, therefore, do not affect $\eta_t$.

For the ideal cogeneration cycle (Figure 7-1), $\eta_t$ is the ratio of $N + Q_t$, which is represented by the area 1a-1-2-3-4-5a, to $Q_f$, which is represented by the same area. Therefore, it can be concluded that for the ideal cycle, $\eta_t$ is always equal to 1.0. By comparing the $\eta_t$ of two plants with similar turbine inlet parameters, but different extraction pressure, it can be concluded that these plants are equally efficient. Is this true?

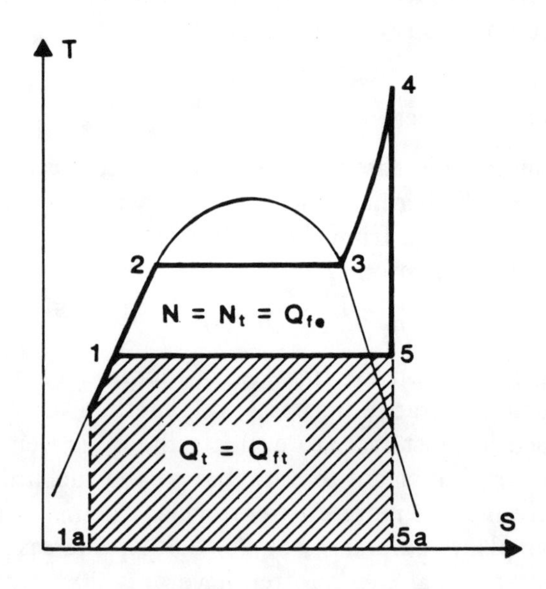


**Figure 7-1. T-S Diagram for Ideal Cogeneration Cycle with Backpressure Steam Turbine**

Table 7-1, Part A, compares two cogeneration plants that utilize controlled back pressure steam turbines: the conventional steam plant and the combined cycle power plant with supplementary firing. Both plants have identical steam turbine inlet parameters and supply the same amount of steam at the same pressure for thermal consumption.

From this table, it follows that the overall thermal efficiency of the combined cycle plant is significantly lower than that of the conventional plant. The air-to-fuel ratio for gas turbines is about three times that of fired boilers; therefore, the generation of equal amounts of electricity in a combined cycle is associated with two and a half to three times higher exhaust gas losses, which shows up in $\eta_t$.

Another method, frequently used for the evaluation of cogeneration effectiveness (also based on the first law of thermodynamics), uses the so-called electrical efficiency (or incremental electric efficiency) of the cogeneration plant ($\eta_{et}$), which can be defined as follows:

$$\eta_{et} = N/(Q_f - Q_t/\eta_b) \tag{7-2}$$

where $Q_t = 0$, $\eta_{et} = \eta_{ec}$ (thermal efficiency of the condensing power plant).

$Q_{ft} = Q_t/\eta_b$ is the amount of fuel required to produce $Q_t$ by the fired boiler, and $Q_{fe}$ is the amount of fuel associated with electrical generation.

$$Q_{fe} = Q_f - (Q_t/\eta_b) = Q_f - Q_{ft} \tag{7-3}$$

$$\eta_{et} = N/Q_{fe} \tag{7-4}$$

It should be emphasized that this method divides the total fuel input into two parts: $Q_{fe}$, for fuel used to produce electrical energy, and $Q_{ft}$, for thermal energy. All the benefits from cogeneration in this method are attributed to electricity generation while the amount of fuel attributed to thermal production is the same as used for a low-potential thermal production plant. According to Figure 7-1, it is apparent that for the ideal cogeneration cycle utilizing back pressure turbines, $N_t$ and $Q_{fe}$, are both represented by the same area, 1-2-3-4-5; therefore, $\eta_{et} = 1$. Like $\eta_t$, $\eta_{et}$ is not affected by the steam inlet and exhaust parameters and turbine internal efficiency.

**Table 7-1**

| *Characteristics* | *Conventional Steam Plant* | *Combined Cycle Plant* |
|---|---|---|
| | *PART A* | |
| 1. Number of Gas Turbines | | |
| 2. Number and Type of Steam Turbines | 1/back pressure | 1/back pressure |
| 3. Live Steam Parameters, psia/°F | 1250/950 | 1250/950 |
| 4. Live Steam Flow, lb/hr | 350,000 | 350,000 |
| 5. Steam Turbine Back Pressure, psia | 400 | 400 |
| 6. Turbine Exhaust Steam Temperature/ Enthalpy, °F/Btu/lb | 690/1358 | 690/1358 |
| 7. Return Condensate Flow/Temperature/ Enthalpy, lb/hr/°F/Btu/lb | 350,000/200/168 | 350,000/200/168 |
| 8. Thermal Energy Supplied (including 5% on-site thermal losses) $(Q_t)$, Btu/hr | $395.5 \times 10^6$ | $395.5 \times 10^6$ |
| 9. Net Power Output $(N)$, kw | 10,420 | 78,320 |
| Including: a) Gas Turbine, kw | – | 68,200 |
| b) Steam Turbine, kw | 10,420 | 10,120 |
| 10. Fuel Input $(Q_f)$, Btu/hr | $505.2 \times 10^6$ | $873.0 \times 10^6$ |
| Including: a) Gas Turbine, Btu/hr | – | $757.8 \times 10^6$ |
| b) Steam Generator, Btu/hr | $505.9 \times 10^6$ | $115.2 \times 10^6$ |
| 11. Fuel Utilization Factor, or Overall Thermal Efficiency $(\eta_t)$ | 0.852 | 0.764 |
| | *PART B* | |
| 12. Fuel Flow for Thermal Energy Production $(Q_{ft})$, Btu/hr | $462.0 \times 10^6$ | $462.0 \times 10^6$ |
| 13. Fuel Flow for Electricity Generation $(Q_{fe})$, Btu/hr | $43.9 \times 10^6$ | $411 \times 10^6$ |
| 14. Plant Net Electrical Efficiency $(\eta_{et})$ | 0.810 | 0.651 |
| | *PART C* | |
| 15. Cogeneration Electricity Output $(N_t)$, kw | 10,420 | 78.320 |
| 16. Power to Heat Ratio $(E_t)$ | 0.09 | 0.675 |
| 17. Difference in Power Output $(\Delta N)$, kw | – | 67,900 |
| 18. Difference in Fuel Input $(\Delta Q_f)$, Btu/hr | – | $367.8 \times 10^6$ |
| 19. Fuel Savings $(\Delta Q_f)$ Comparison of Non-Cogeneration (separate 9000 Btu/kwh Condensing Plant and Industrial-Steam Boilers), Btu/hr | $49.8 \times 10^6$ | $293.9 \times 10^6$ |
| 20. Relative Fuel Savings $(\delta Q_f)$, % | 9.00 | 25.2 |

Also, like $N_t$, only losses that are related to electricity generation and are not later recovered as thermal energy actually affect $\eta_{et}$. For these reasons, $\eta_{et}$ for the conventional steam plant is higher than it is for the combined cycle plant as shown in Table 7-1, Part B, which illustrates the weakness of this method of evaluating cogeneration.

As will be shown later, the combined cycle plant is a more efficient cogeneration source than the conventional steam plant. Additionally, plants that have higher inlet steam parameters and/or lower steam turbine back pressures are more efficient than plants with lower inlet steam parameters and higher back pressures.

## Electrical and Thermal Energy in the Cogeneration Schemes

As can be seen in equations (7-1) and (7-2) the parameters $\eta_t$ and $\eta_{et}$ do not separate the two types of energy, electrical and thermal, and they are treated equally. Electricity is of a higher potential energy; its production is normally associated with much greater losses than the production of thermal energy.

According to the second law of thermodynamics, thermal energy cannot be transformed into work without rejecting some amount of energy from the cycle. Even the most efficient power plants convert less than 45% of the fuel energy into electricity with about 50% of the total energy rejected to the atmosphere as very low-potential thermal energy.

In the meantime, conversion of fuel energy into heat does not require energy rejection from the cycle. This means that fuel energy can be totally converted to thermal energy except for boiler and other minor losses.

The major difference between electrical and thermal energy lies in their ability to do work. Electrical energy can be almost entirely converted into mechanical energy, but only part of the thermal energy can be converted. Thermal energy of lower potential (i.e., different pressure and temperature) produces less mechanical work.

A major factor that defines the importance or worth of energy is its ability to produce a certain amount of work. Based on this factor, electricity is normally worth two and a half to three times more than low-potential thermal energy. *It can be concluded that electrical and thermal energy are not equal and, therefore, cannot be treated equally.*

Cogeneration cycles are not in contradiction with the second law of thermodynamics because the rejected heat of low-potential thermal energy is used later. At the expense of electrical energy, thermal energy is extracted at a great enough potential that it can be used as a source of heat or can even produce some mechanical work; but for electrical generation, this rejected heat is a loss. *The real thermodynamic advantage of cogeneration is that the thermal energy is supplied as a by-product of the electrical generation cycle* rather than being produced from full potential fuel energy; for example, thermal energy is supplied as low potential extracted steam.

Because the turbines for cogeneration are generally smaller and designed for simpler operation, the thermal efficiency of the utility power plant normally is higher than the thermal efficiency of the cogeneration plant (when no useful thermal energy is produced).

On the other hand, cogenerated electricity is produced with virtually no thermal loss, whereas this loss is approximately 50% for condensing power plants. As a result of this, the maximum effectiveness of a cogeneration plant will be achieved when all of the electricity is generated by steam later utilized as thermal energy.

In the overwhelming majority of cases, it is economically advantageous to generate as much electrical power as possible. If the amount of electricity generated by the cogeneration plant exceeds the requirements of local customers, the additional electricity may be used to replace the more expensive energy generated by a utility plant. *Only the electrical power generated by steam (or another working medium such as combustion gases) that is later used as useful thermal energy realizes a fuel savings.*

## Alternative Evaluation Methods

The overall energy effectiveness of cogeneration can be estimated by comparing the cost of fuel used in the cogeneration cycle to produce the required amounts of thermal and electrical energy with the cost of fuel and electricity necessary to produce the same amounts of both types of energy when produced by alternative sources (i.e., condensing power plants, industrial or heating boilers, etc.). The cost of purchased electricity is determined primarily by the amount and cost of fuel used to produce this electricity.

Usually, the cost for the same fuel will not vary substantially between the utility plant and cogeneration plant. Therefore, the overall effectiveness of cogeneration is a function of the fuel effectiveness of cogeneration, which in turn can be defined as the difference between the amount of fuel used in the cogeneration cycle and the amount of fuel used in a noncogenerated scheme. This can be expressed by the following equations:

Absolute fuel energy savings:

$$\Delta Q_f = Q_{fncg} - Q_{fcg} \quad (7\text{-}5)$$

Relative fuel energy savings:

$$\delta Q_f = (Q_{fncg} - Q_{fcg})/Q_{fncg} \quad (7\text{-}6)$$

In most cases: $Q_{fncg} = Q_{fcn} + Q_{fb}$

If $\Delta Q_f > 0$, there is a fuel savings; if $\Delta Q_f < 0$, the cogeneration plant is thermally less efficient than the alternative sources of energy. The larger the $\Delta Q_f$, the more effective the cogeneration plant. Therefore, as $\Delta Q_f$ increases, so does the fuel effectiveness of the cogeneration plant.

If the utility plant uses less expensive fuel than the cogeneration plant, the difference in fuel energy consumption $(\Delta Q_f)$ needs to be sufficiently high to offset the higher fuel costs.

The amount of thermal energy provided by the cogeneration plant $(Q_f)$ is fixed by the process requirements and closely corresponds to the amount of cogeneration electricity produced $(N_t)$. Let's define

$$N_t/Q_t = E_t, \quad (7\text{-}7)$$

where $E_t$ {Btu/Btu} = 293 $E_t$ {kwh/$10^6$ Btu}. $E_t$ is the ratio of electricity generation to the thermal output *in the cogeneration process.* Also $E_t$ can be defined as the amount of electricity generated for each $10^6$ Btu/hr (or lb/hr of steam) of useful thermal energy produced.

The power-to-heat ratio $(E_t)$ varies greatly for different types of cogeneration schemes. The steam turbine topping cycle has the lowest power-to-heat ratio, between 20 and 60 kwh/$10^6$ Btu/hr, while the combined-cycle plants have a power-to-heat ratio of between 200 and 450 kwh/$10^6$ Btu. As a result, the combined cycle-plant normally produces several times more electricity than the equivalent (in the thermal sense) steam turbine plant.

A simplified comparison of the fuel effectiveness of two cogeneration plants (plant 1 and plant 2), both supplying the same $Q_t$ but generating different amounts of cogeneration electricity, is shown below:

Fuel savings in each case compared to the noncogeneration case are:

$$\Delta Q_1 f = (N_1 t/\eta ec + Q_t/\eta b) - Q_1 f \tag{7-8}$$

$$\Delta Q_2 f = (N_2 t/\eta ec - Q_t/\eta b) - Q_2 f \tag{7-9}$$

From equations (7-3) and (7-4) it follows that

$$Qf = Qfe + Q_t/\eta b = N_t/\eta et + Q_t/\eta b \tag{7-10}$$

By replacing $Q_1 f$ and $Q_2 f$ in equations (7-8) and (7-9) it follows that

$$\begin{aligned}\Delta Q_1 f &= N_1 t/\eta ec + Q_t/\eta b - N_1 t/\eta_1 et - Q_t/\eta b \\ &= E_1 t Q_t (1/\eta ec - 1/\eta_1 et)\end{aligned} \tag{7-11}$$

$$\Delta Q_2 f = N_2 t(1/\eta ec - 1/\eta_2 et) = E_2 t Q_t (1/\eta ec - 1/\eta_2 et) \tag{7-12}$$

From this simplified analysis, it follows that the fuel savings is proportional to the amount of cogenerated electricity produced and the difference between the electrical heat rates of the utility station and the cogeneration plant. Also, it follows that the generation of electricity in the cogeneration process is the key factor affecting the fuel effectiveness of cogeneration. Generation of electricity by the cogeneration plant in a noncogeneration process (i.e., condensing) normally does not result in any appreciable fuel savings. (In fact, there is often a loss.)

Since $N_t$ and $E_t$ are in a fixed relationship, the increase in $E_t$ brings additional savings for cogeneration schemes. Equations (7-11) and (7-12) also illustrate that the cogeneration plant with the lower $\eta_{et}$ (i.e., combined cycle vs. steam turbine topping cycle) may have significantly greater fuel savings (due to the $E_t$, which is several times larger).

This method, as used for an analysis of a combined cycle and conventional steam plant, is illustrated in Table 7-1 Part C.

Figure 7-2 illustrates why a greater power-to-heat ratio in the cogeneration cycle results in higher cogeneration effectiveness. As we can see from this figure, the overall fuel utilization factor in the noncogeneration case (as represented by the solid line) is decreasing from about 0.82 for strictly industrial boilers with power-to-heat ratio ($E_t$) equal to zero (with no electricity use) to less than 0.4 for a very large $E_t$ (central power station and very little or no useful thermal energy). The shaded areas represent a typical F.U.F. range for various types of cogeneration plants at various $E_t$. The vertical distance between cogeneration plant F.U.F. and noncogeneration F.U.F. represents actual relative fuel savings at various power-to-heat ratios. As an example, the data from Table 7-1 are plotted in Figure 7-2.

Lastly, we would like to evaluate the sensitivity of cogeneration plant fuel saving from the standpoint of various efficiency factors. Let us define cogeneration plant relative fuel profit as:

$$\Pi Q_{cg} = (Q_{fncg} - Q_{fcg})/Q_{fcg} \qquad (7\text{-}13)$$

Combining equations (7-13), (7-8) and (7-2) the following can be obtained:

$$\Pi Q_{cs} = [N_t/\eta_{ec} + Q_t/\eta_b - (N_t + Q_t/\eta_t)] \div (N_t + Q_t)/\eta_b \qquad (7\text{-}14)$$

Simplifying equation (7-14) it follows that:

$$\begin{aligned}\Pi Q_{cg} &= \eta_{ecg}/\eta_{ec} + (\eta_t - \eta_{ecg})/\eta_b - 1\\ &= \eta_{ecg}\,(1/\eta_{ec} - 1/\eta_b) + \eta_t/\eta_b - 1\end{aligned} \qquad (7\text{-}15)$$

where

$$\eta_{ecg} = N_t\,\eta_t/(N_t + Q_t) \qquad (7\text{-}16)$$

$\eta_{ecg}$ can be defined as cogeneration plant overall electrical efficiency based on total fuel input. The principal difference between $\eta_{et}$ and $\eta_{ecg}$ is that in calculating $\eta_{ecg}$ no credit is given for useful thermal energy production.

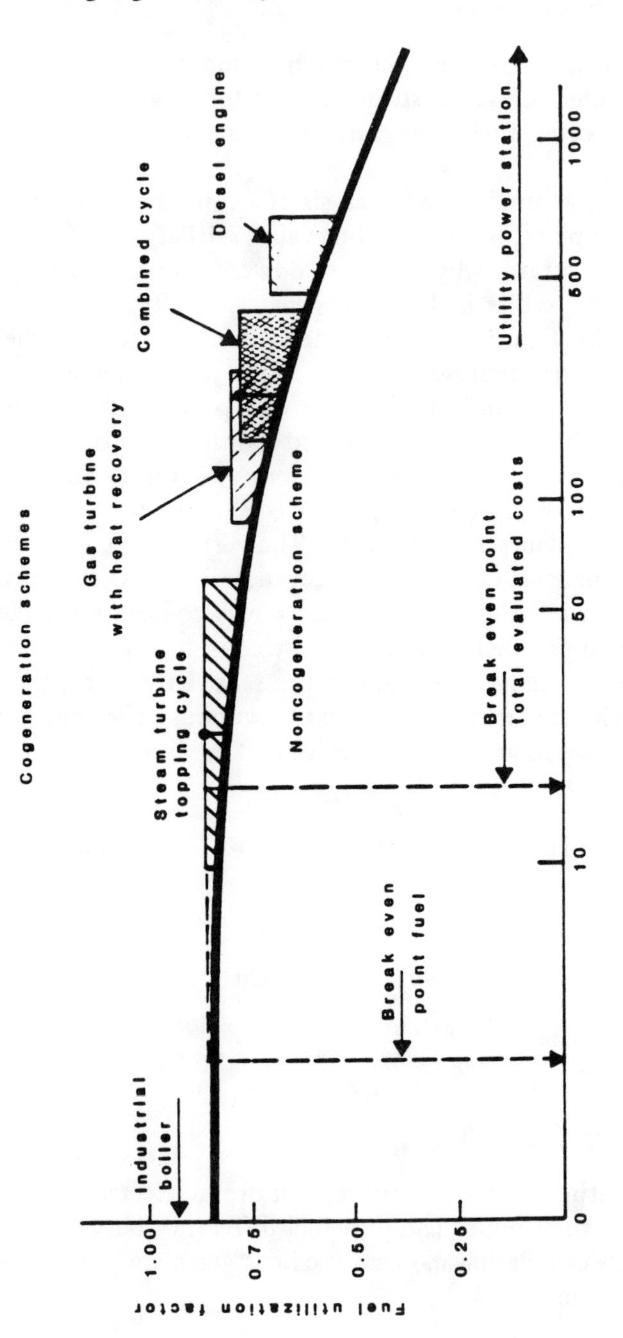


Figure 7-2. Fuel Utilization for Various Types of Installations

Replacing $\eta_{ecg}$ in equation (7-15) by the electrical heat rate, H. R. = $3413/\eta_{ecg}$, it follows that

$$\Pi Q_{cg} = (3413/\text{H.R.})(1/\eta_{ec} - 1/\eta_b) + \eta_t/\eta_b - 1 \qquad (7\text{-}17)$$

Analyzing equations (7-15) to (7-17) we can obtain very interesting results:

1. The cogeneration plant relative fuel profit (which is almost all operating profit) is a linear function of cogeneration plant overall electrical efficiency ($\eta_{ecg}$) and overall thermal efficiency ($\eta_t$).
2. Relative fuel profit is also a function of the utility electrical efficiency ($\eta_{ec}$) and the industrial boiler thermal efficiency ($\eta_b$). For a given cogeneration plant $\eta_{ec}$ and $\eta_b$ normally are constant.

Efficiency $\eta_{ecg}$ is a function of the type of cogeneration plant (Diesel, combined cycle, steam turbine, topping, etc.), while $\eta_t$ is a function of the cycle energy losses.

$\eta_t$ represents the efficiency with which the overall fuel energy is converted to both thermal and electrical energy regardless of the quantity of each type of energy. Equation (7-15) helps to explain one very important fallacy of a traditional cogeneration analysis. It is quite common to hear that the electrical efficiency of the prime mover used in a cogeneration plant is unimportant because all thermal energy is later recovered in the form of useful thermal energy. Now we can see that this conception is totally incorrect. From equation (7-15) it follows that for two cogeneration plants of equal thermal efficiency (identical $\eta_t$), the plant with the higher overall electrical efficiency ($\eta_{ecg}$) produces the higher energy (fuel) saving. For example, based on data from Table 7-1 where $\eta_{ec} = 37.9\%$ and $\eta_b = 0.82$,

$$\Pi Q_{eg} = 1.417\ \eta_{ecg} - 1.22\ \eta_t - 1 \qquad (7\text{-}18)$$

Figure 7-3 is a graphical representation of equation (7-18) for various $\eta_{ecg}$ and $\eta_t$. For the cogeneration plants shown in Table 7-1 $\Pi Q_{cg}$ = .14 for a conventional steam plant and 0.36 for a combined cycle. From Figure 7-2 it follows that the higher thermal efficiency

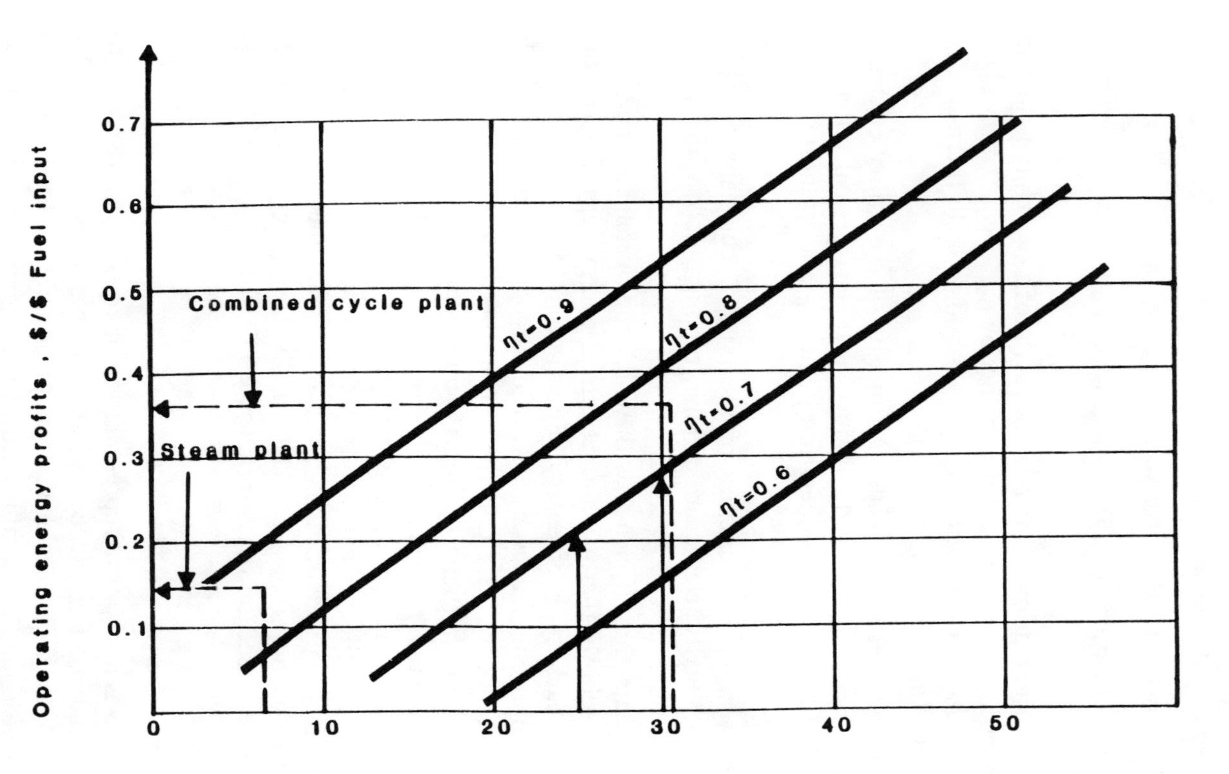


Figure 7-3. Cogeneration Plant Specific Operating Energy Profits vs. Plant Efficiency

of the prime mover would result in higher cogeneration profits. For example, a 5% increase in gas turbine efficiency (using a 30% efficiency gas turbine versus a 25% efficient one and $\eta_t = 0.7$) could improve relative cogeneration profits from 0.21 to 0.28 or by 33%.

**Summary**

- To correctly judge the effectiveness of cogeneration, the meaning of cogeneration effectiveness must be clearly understood, and proper methods of evaluating it must be used.
- Overall cycle thermal efficiency or fuel utilization factor $(\eta_t)$ and cogeneration cycle electrical efficiency $(\eta_{et})$ cannot be used in most cases to evaluate the effectiveness of cogeneration.
- Electrical and thermal energy generated by the cogeneration plant are not of equal quality and, therefore, cannot be treated equally. The generation of electricity as a by-product of thermal energy production in the cogeneration schemes is the key factor affecting the fuel effectiveness of cogeneration. Therefore, evaluation methods that are based on the ratio of electrical to thermal energy production are most suitable for evaluating the effectiveness of cogeneration.
- A higher power-to-heat ratio normally results in higher fuel savings and, therefore, is more economically attractive.
- Prime mover electrical efficiency has a substantial effect on cogeneration profitability. Higher efficiency results in much greater fuel savings and profits despite the fact that overall thermal efficiency remains unchanged.

**Nomenclature**

$N$ total amount of electricity generated
$N_t$ cogenerated electricity
$N_c$ condensing electricity
$Q_t$ thermal energy supplied from the cogeneration cycle
$Q_f$ fuel input
$Q_{ft}$ the amount of fuel associated with thermal energy production

| | |
|---|---|
| $Qfe$ | the amount of fuel associated with electricity generation |
| $Qfncg$ | fuel used by noncogeneration sources of energy |
| $Qfcg$ | fuel used by the cogeneration plant |
| $Qfcn$ | fuel used by a competitive condensing plant to generate electricity |
| $Qfb$ | fuel used to generate thermal energy by the plant other than cogeneration (i.e., industrial boilers) |
| $\eta t$ | overall plant thermal efficiency or fuel utilization factor (F.U.F.) |
| $\eta et$ | electrical efficiency of the cogeneration plant |
| $\eta b$ | steam generator thermal efficiency |
| $\eta ec$ | thermal efficiency of the condensing plant |
| $\eta ecg$ | cogeneration plant overall electrical efficiency |

# Chapter 8

# Industrial Cogeneration: System Application Considerations

***J. M. Kovacik***

Cogeneration has been practiced by many industries during this century as a reliable, economic means of generating power in conjunction with satisfying process heating needs. Prior to the 1960s, most cogeneration systems were based on the use of steam turbine-generators in both non-extraction and automatic-extraction designs. Most applications were associated with pulp and paper, textile, chemical and food industries.

During the 1960s, industrial managers recognized the economic benefits that could be realized through use of gas turbine cogeneration systems; and, where suitable fuels were economically available, gas turbine cogeneration applications flourished. This prime mover has maintained a favorable reputation as a critical element contributing to effective cogeneration systems. Furthermore, recent legislation has stimulated interest and provided additional potential for gas turbine cogeneration systems in the years ahead.

This chapter reviews application considerations for both steam turbine and gas turbine cogeneration systems. The impact of the Public Utility Regulatory Policies Act (PURPA) on the development of alternatives that merit consideration is briefly discussed.

## Cogeneration

Cogeneration has often been defined as the sequential production of useful thermal energy and shaft power from a single energy source. The shaft power can be used to drive either mechanical equipment such as pumps and compressors, or electric generators. Power produced in the manner described is called a "topping" cogeneration cycle. (A "bottoming" cogeneration system produces power resulting from low level energy recovery associated with the process).

The more effective use of energy in topping cogeneration systems is illustrated in Figure 8-1. The power generation case is typical of what can be expected from a modern coal-fired power generating plant. The 35% energy utilization is equivalent to a 9750 Btu/kwh HHV (higher heat value) heat rate for this coal-fired system.

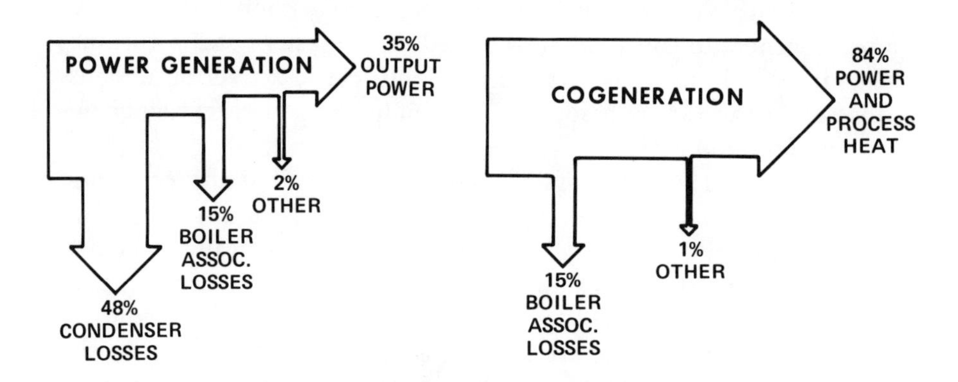


**Figure 8-1. Fuel Utilization Effectiveness**

The "cogeneration" arrow in Figure 8-1 illustrates the higher energy utilization effectiveness realized when heat and power are delivered from a single system. In a topping cogeneration system, the industrial process becomes the heat sink (or condenser) for the power generation system. Thus, the cogeneration system can often deliver 80% or more of the fuel input energy as useful output (heat and power) for plant use.

## Steam Turbine Cycles

Steam turbine cycles are frequently applied in those industrial plants having process by-products available as the cogeneration system fuel. These by-product fuels include black liquor and hog fuel in the pulp and paper industry, wastes associated with the food industry, blast furnace and coke oven gas from steel mill operations, and others. Also, in those applications where coal or residual fuel oil are the economic plant fuels, steam turbine cycles are usually preferred relative to other topping cogeneration cycle options.

Development of economic steam turbine cogeneration systems requires careful evaluation of the following aspects of the system design:

- Prime mover size
- Initial steam conditions
- Feedwater heating
- Condensing power

### *Prime Mover Size*

The application of more efficient large steam turbine-generators rather than a group of smaller less efficient mechanical drive steam turbines as drivers for small power plant mechanical loads, can significantly improve energy utilization. The example given in Table 8-1 shows that 70 percent more power can be generated expanding the steam required for process in the larger turbine-generator rather than the smaller mechanical drive units. Considering the losses associated with the transformation and distribution of energy from the turbine-generator to the motors for the small plant auxiliaries, the net gain is approximately 57 percent rather than the 70 percent implied by the data in Table 8-1.

For grassroots facilities, major industrial power plant expansions, or modernization programs, economics usually favor expansion of steam in a steam turbine-generator and use of motors rather than turbines to drive small mechanical loads. However, replacement of existing noncondensing turbines driving miscellaneous loads is more difficult to economically justify since the capital burden of the displaced capacity must be recovered through the annual savings result-

**Table 8-1. Influence of Prime Mover Size on Cogenerated Power**

| *Type Prime Mover* | *Approximate Efficiency* | *Cogenerated Power per 100 Million Btu/hr Net Heat to Process* |
|---|---|---|
| 500 hp Single Stage Mechanical Drive Units | 45% | 2800 kw eq. |
| 5000 kw Multi-Valve Multi-Stage Turbine-Generator | 72% | 4770 kw |

*Basis:*

1) Initial steam conditions, 600 psig, 750F.
2) Process steam required at 50 psig.
3) Process returns at 180F.
4) No feedwater heating has been included.

ing from the effective increase in cogenerated power. For example, based on Table 8-1 conditions, the incremental capital cost (installed) of the 5000 kw turbine-generator, the replacement motors, and the associated electrical equipment relative to the mechanical drive case would have to be justified based on the 1600 kw incremental increase in power generated.

## *Initial Steam Conditions*

The increased thermodynamic availability of steam at various initial steam conditions and exhaust pressures relative to 600 psig, 750F steam is illustrated in Figure 8-2. The data presented are based on initial steam temperatures that will provide essentially the same expansion line at a 75 percent turbine efficiency regardless of the initial pressure selected. The data illustrate that the magnitude of the gain in cogenerated power that can be realized is a function of the initial steam conditions selected as well as the pressure level or levels required in process.

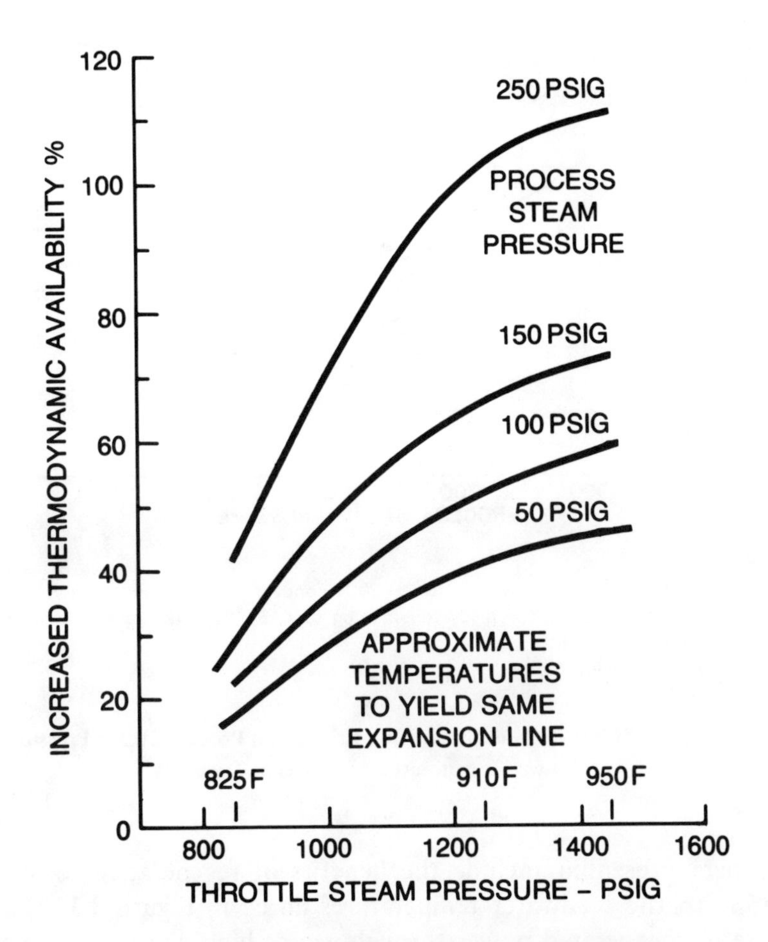


**Figure 8-2. Increased Thermodynamic Availability for Various Throttle Steam Conditions**

The effect of turbine inlet steam conditions on the amount of power that can be cogenerated per 100 million Btu/hr net heat to process (NHP)* at different steam pressures is shown in Figure 8-3.

*Net heat to process is the net energy delivered to the process without consideration of the boiler inefficiency.

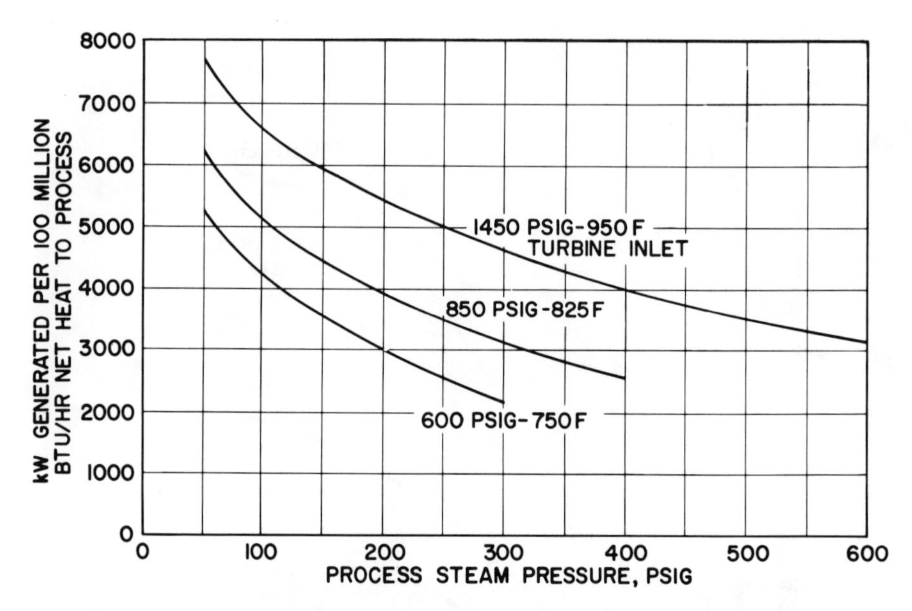


*Basis:*

1) 70% of process steam flow returned as 200F condensate and balance as 80F makeup.
2) Power cycle credited for feedwater heating to 445F for 1450 psig unit, 400F for 850 psig, 370F for 600 psig.
3) Turbine efficiency 75%.

**Figure 8-3. Effect of Inlet Steam Conditions and Process Steam Pressure on Power Cogenerated with Steam Turbines**

The data presented include the benefits of regenerative feedwater heating to the feedwater temperatures noted in Figure 8-3. The increase in cogenerated power through use of higher initial steam conditions is readily apparent. (Figure 8-3 also illustrates the output gains that are forfeited if plants are designed with excessive margin in the process steam distribution systems.)

Studies have shown that the higher steam conditions can be more easily economically justified in industrial plants having relatively large process steam demands. Data given in Figure 8-4 provide guidance with regard to the initial steam conditions that are normally considered for industrial cogeneration applications. Higher energy

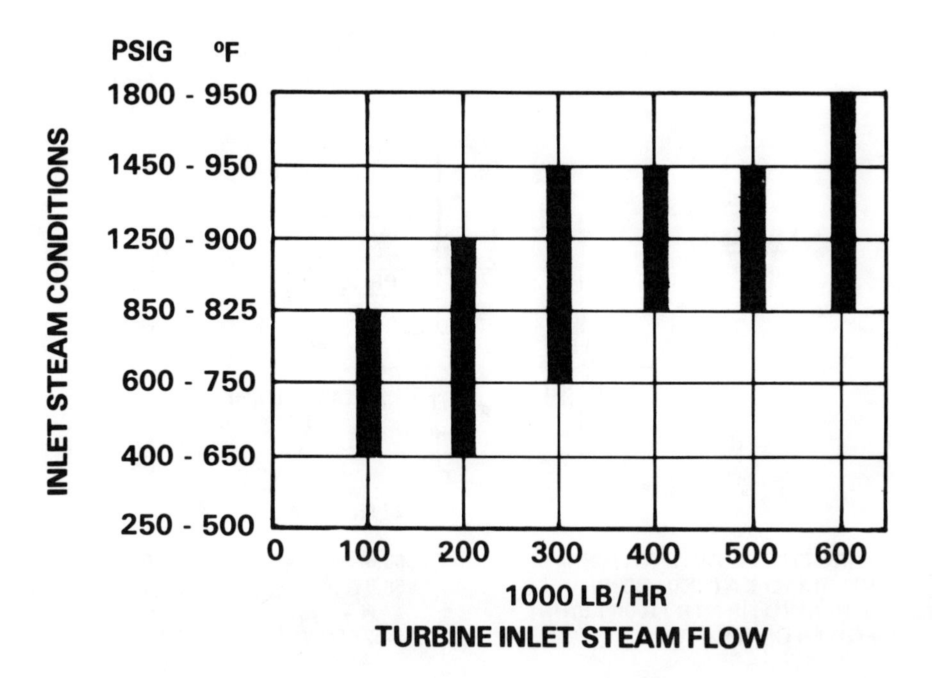


**Figure 8-4. Range of Initial Steam Conditions Normally Selected for Industrial Steam Turbines**

costs experienced since the mid 1970s are favoring the upper portion of the bands shown in Figure 8-4.

*Feedwater Heating*

Feedwater heating provides a means of increasing cycle steam requirements and thus the amount of power that can be cogenerated in a steam turbine cycle. The simple example given in Figure 8-5 shows that the addition of the closed heater at 225 psig increases the cycle steam needs about 10 percent and the amount of power that can be cogenerated by 8.3 percent.

Since many industrial plants have several process pressure levels, the individual process pressures may be logical locations for feedwater heaters. Generally speaking, the use of three stages of feedwater heating is economic for most industrial applications using steam turbine cogeneration cycles.

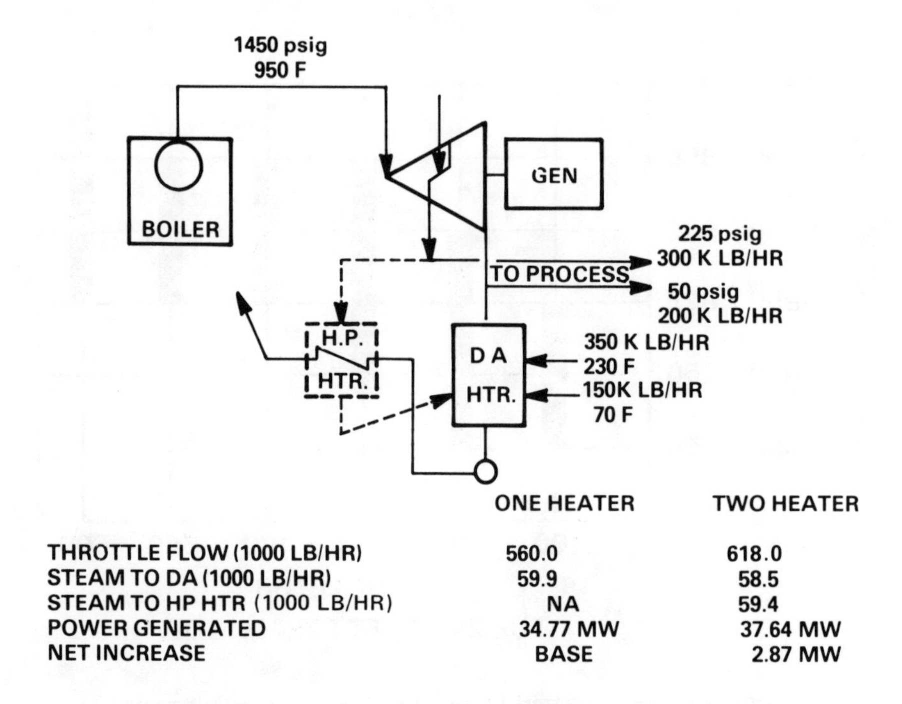


| | ONE HEATER | TWO HEATER |
|---|---|---|
| THROTTLE FLOW (1000 LB/HR) | 560.0 | 618.0 |
| STEAM TO DA (1000 LB/HR) | 59.9 | 58.5 |
| STEAM TO HP HTR (1000 LB/HR) | NA | 59.4 |
| POWER GENERATED | 34.77 MW | 37.64 MW |
| NET INCREASE | BASE | 2.87 MW |

**Figure 8-5. Effect of Added Feedwater Heating on Cogenerated Steam Turbine Power**

*Condensing Power Generation*

The amount of power that can be cogenerated in steam turbine cycles is a function of the initial steam conditions, the process pressure level(s), and the feedwater heating cycle selected. In addition, the data in Figure 8-3 show that even for the most effectively developed systems, it is not likely that the amount of power that is generated per unit of heat to process will exceed 80 kw per million Btu NHP. This is usually less power than required to satisfy most industrial plant electrical energy needs. Thus, condensing power is frequently considered to augment steam turbine cogenerated power.

The impact of adding condensing power generation to a "pure" steam turbine cogeneration system (noncondensing steam turbine power) is illustrated in Figure 8-6. Even though condensing power is

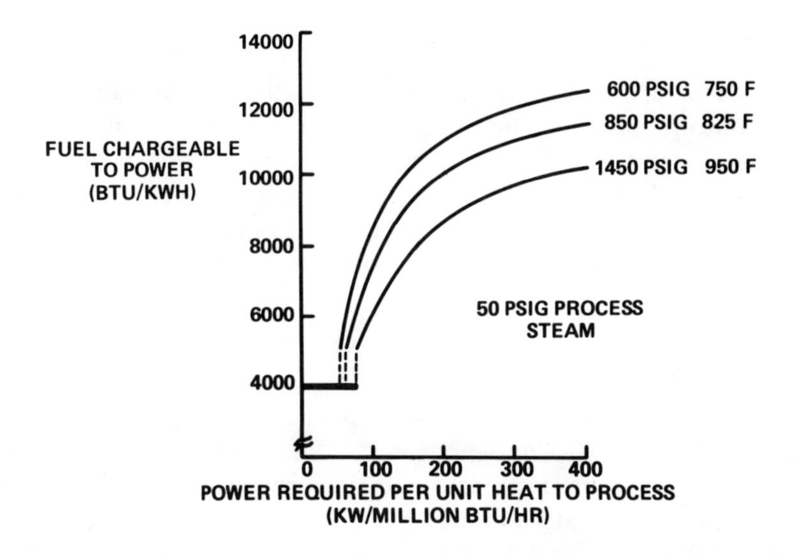


*Basis:*

1) Process steam at 50 psig.
2) Cycle specifics as in Figure 8-3.
3) Fuel chargeable to power (FCP) is the incremental increase in fuel consumption due to the cogeneration system divided by the net power generation credited to the cycle.
4) FPC based on credit for process heat at 84% boiler efficiency.

**Figure 8-6. Fuel Chargeable to Power—Various Steam Turbine Cycles**

not necessarily energy efficient, it can occasionally prove economic. Favorable economics can frequently be realized for the following conditions:

- Condensing power can be used for utility demand control.
- Low cost fuels such as wood or coal, or excess process by-products are available as plant fuels.
- Thermal energy from process is available for expansion to a condenser (bottoming cogenerated power).
- Reliability of other power sources is questionable.

## Gas Turbines and Combined Cycles

Gas turbine cycles provide the opportunity to generate a large power output per unit of heat required in the process relative to non-condensing steam turbine cogeneration systems. This characteristic combined with a favorable fuel chargeable to power (FCP)* and proven reliability is why this prime mover has had wide acceptance in industrial plants where suitable fuels are economically available.

### *Cycle Configurations*

The various cycle options for gas turbines with HRSG's are illustrated in Figure 8-7. Configuration "A" has the HRSG generating steam at the appropriate steam conditions for process use and is the most simple configuration available.

The HRSG in Figure 8-7B generates steam at elevated steam conditions so both steam turbine and gas turbine power can be cogenerated. This configuration yields the highest power to heat ratio for any configuration presented in Figure 8-7.

An arrangement commonly applied as a result of the rapid increase in fuel costs experienced since 1973 is shown in Figure 8-7C. This multiple pressure HRSG is usually applicable when the gas temperature entering the HRSG surface is about 1200F or lower. Multiple pressure HRSG's provide increased recovery of the gas turbine exhaust energy relative to the Figure 8-7B configuration and thus contribute to the favorable FCP associated with these cycles. FCP improvements in the 10 to 15 percent range are typical for these two-pressure-level systems.

Figure 8-7D, an arrangement that includes a condensing section on the steam turbine-generator, is an extension of the Figure 8-7C configuration. The condensing section on the steam turbine provides cycle flexibility permitting utilization of HRSG steam production in excess of process steam demands, but at an increased cycle FCP.

---

*Fuel chargeable to power is the incremental increase in fuel consumption due to cogeneration divided by the net power generation credited to the cycle.

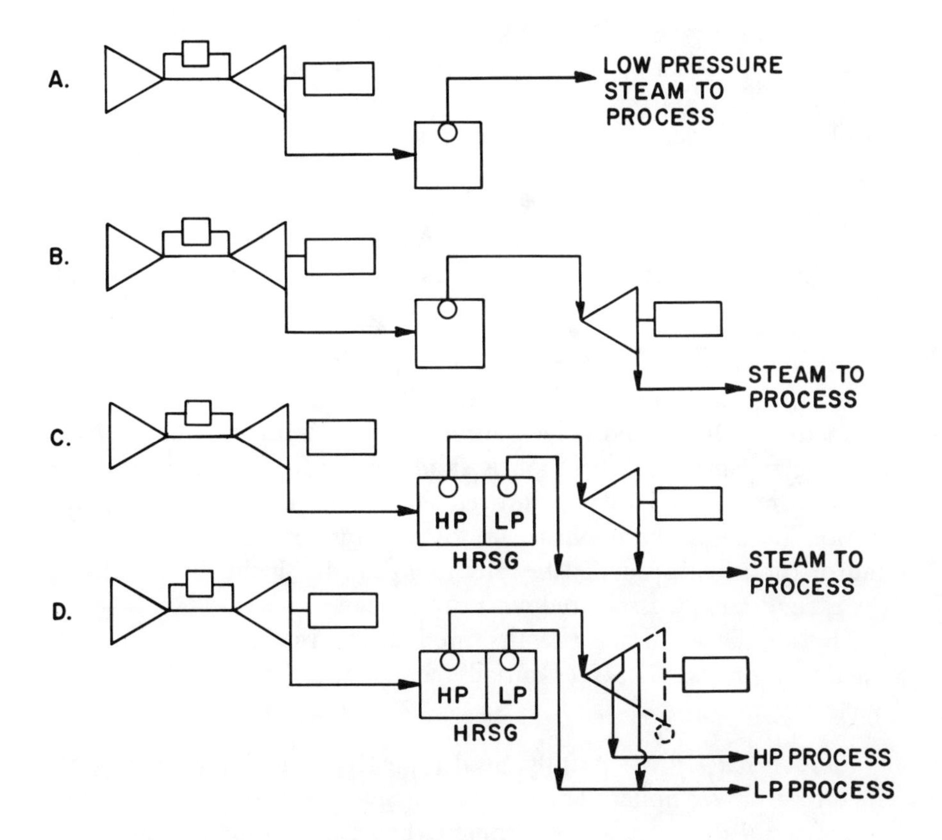


**Figure 8-7. Possible Plant Energy Systems Gas Turbine with Heat Recovery Steam Generators**

*Heat Recovery Steam Generators*

Developers of gas turbine cogeneration systems may have several HRSG options. These include:

- Unfired HRSG's
- Supplementary fired HRSG's
- Fully fired HRSG's

*Unfired HRSG.* An unfired HRSG is an extended surface convective heat exchanger designed to recover a portion of the sensible heat in the gas turbine exhaust. These units can provide steam at low steam conditions such as 150 psig (saturated) for direct use in process. Or, steam can be generated at elevated steam conditions for expansion in a steam turbine prior to delivery to the process. Some gas turbine units have exhaust temperatures of 1000F or somewhat higher, temperatures adequate for generation of steam for use at 1500 psig and 925F, if desired.

The unfired HRSG steam production is a function of the exhaust flow and temperature entering the unit. Thus, the unit is a slave of the gas turbine and cannot be controlled.

*Supplementary Fired HRSG.* In a supplementary fired application, an auxiliary burner is used to increase the turbine exhaust temperature to 1700F or less. These units are essentially convective heat exchangers whose construction is similar to unfired designs. The primary difference relative to unfired HRSG's is in the heat transfer section immediately downstream of the burner where bare tubes and/or tubes with reduced fin pitch and height shield the unit from the radiant energy associated with the burner.

The auxiliary burner permits modulating the steam production capability of the HRSG essentially independent of the gas turbine operating mode.

*Fully Fired HRSG.* A fully fired type HRSG is similar in appearance to a power boiler. The design usually admits to its combustion system only the amount of turbine exhaust gas required to generate the desired amount of steam. The balance of the exhaust flow is bypassed and rejoins the gases used for combustion ahead of the heat recovery section.

The maximum amount of steam that can be generated in a fully fired HRSG is usually 6 to 7 times that available from an unfired HRSG. Also these cycles provide the lowest (best) FCP. Even so, fully fired HRSG's have not been widely applied in industrial applications.

*Estimating Steam Production—Unfired and Supplementary Fired Units.* A simplified diagram illustrating the temperature relationships governing unfired HRSG designs is shown in Figure 8-8. The temperature difference ($T_2-T_3$) is frequently referred to as the "pinch point," and is governed by the effectiveness and the degree of sub-

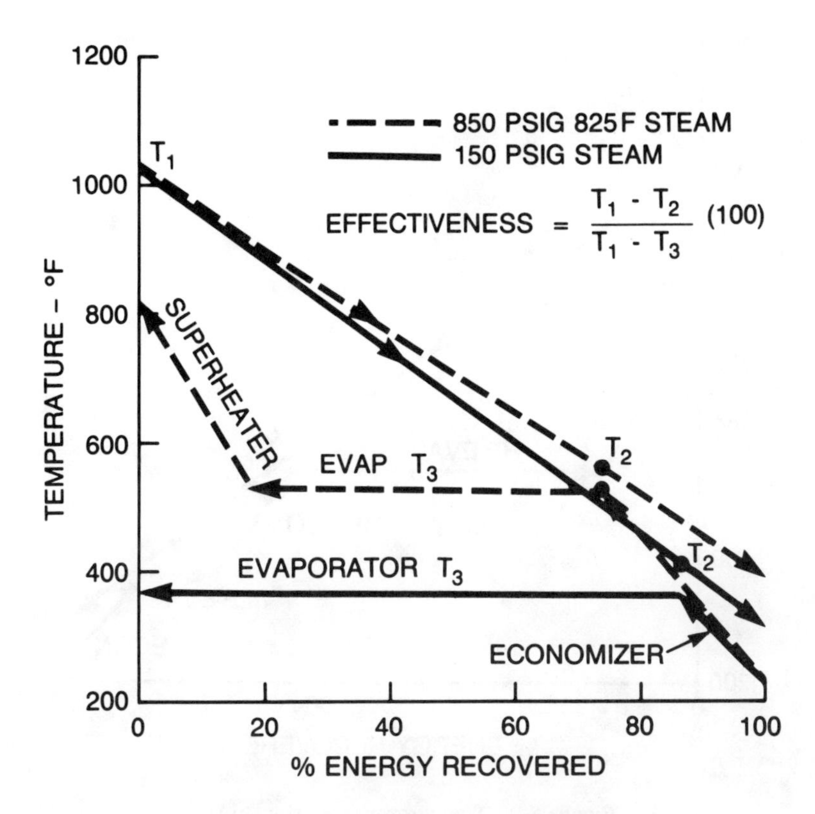


**Figure 8-8. Temperature Relationships Unfired Heat Recovery Boiler**

cooling designed into the economizing section. Figure 8-8 also shows the importance of using a low feedwater heating temperature in gas turbine HRSG systems which is in contrast to steam turbine cogeneration systems where a larger amount of feedwater heating is usually desirable.

The data presented in Figure 8-8 also illustrate the lower energy recovery if an HRSG is designed to provide higher steam conditions required to support combined cycles. However, in these instances, opportunities for multiple levels of energy recovery as shown in Figure 8-9 can provide the added benefits of steam turbine cogeneration and a reasonable stack temperature.

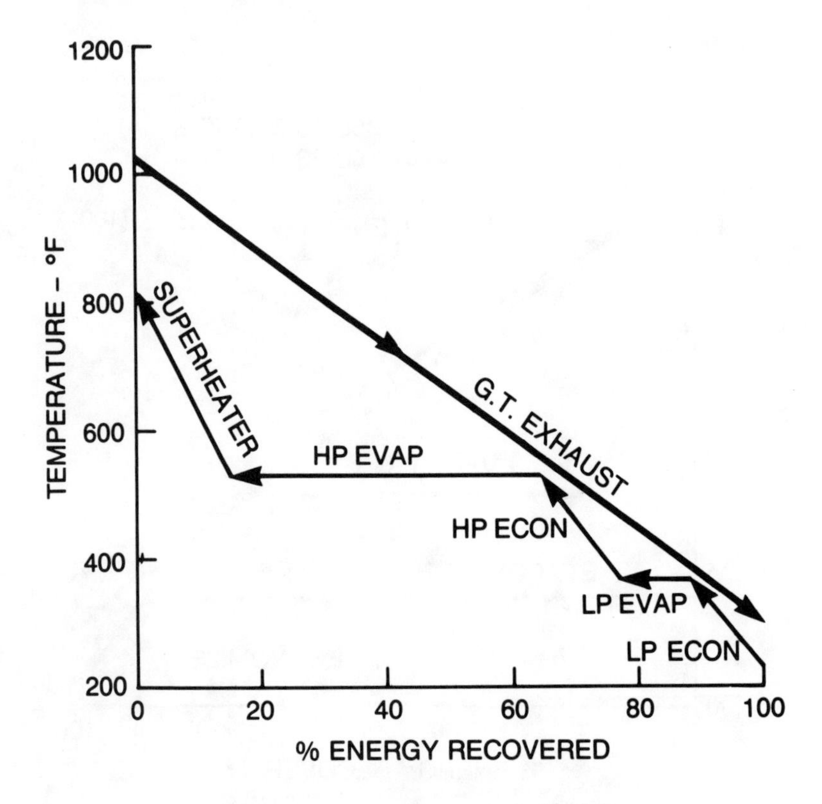


**Figure 8-9. Temperature Relationships Two-Pressure-Level Unfired Heat Recovery Boiler**

The amount of steam that can be generated in single pressure unfired or supplementary fired HRSG's can be estimated using the following relationship:

$$W_{stm} = \frac{W_{exh}}{10^6} \cdot F_1 \cdot F_2$$

Where: $W_{stm}$ is the steam generated

$W_{exh}$ is the gas turbine exhaust flow in lb/hr

$F_1$ is the saturated steam production based on the steam pressure desired and the gas temperature entering the heat transfer surface (see Figure 8-10)

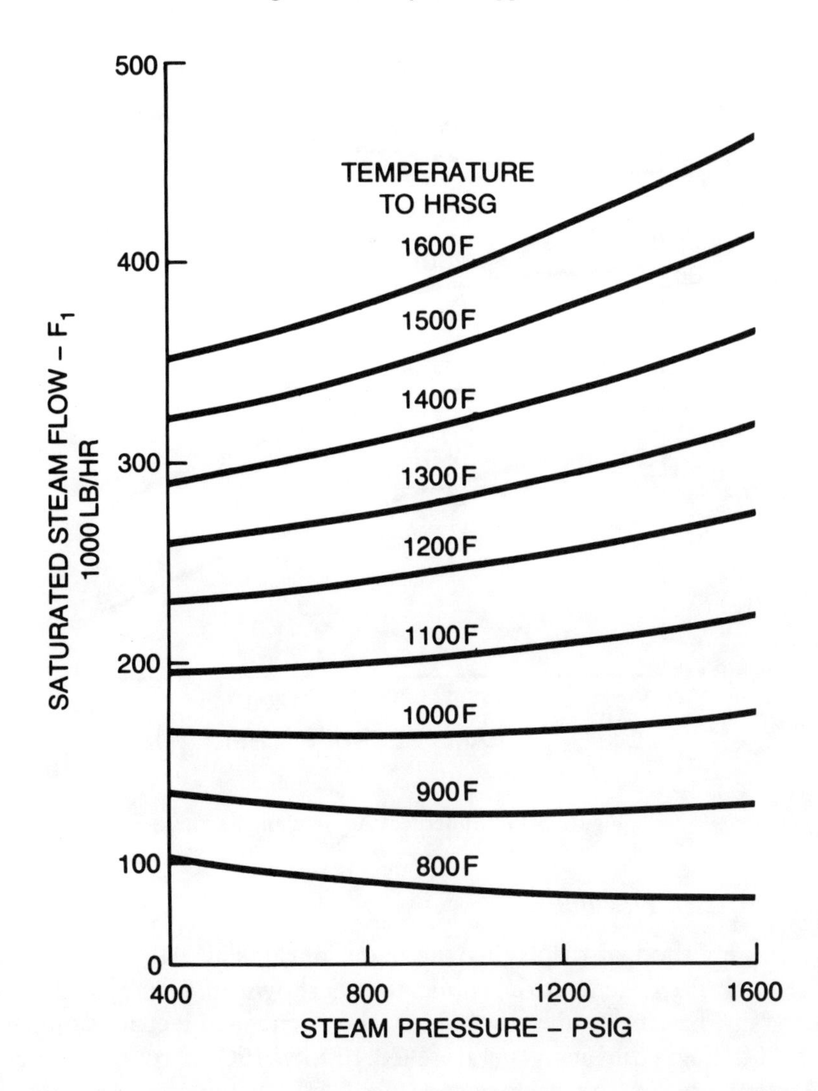


**Figure 8-10. Saturated Steam Production in Heat Recovery System Generators**

$F_2$ is a factor that adjusts the HRSG production to the desired steam temperature (see Figure 8-11)

For units fired to average exhaust gas temperatures of 1700F or less, the HRSG fuel requirements can be estimated using Figure 8-12.

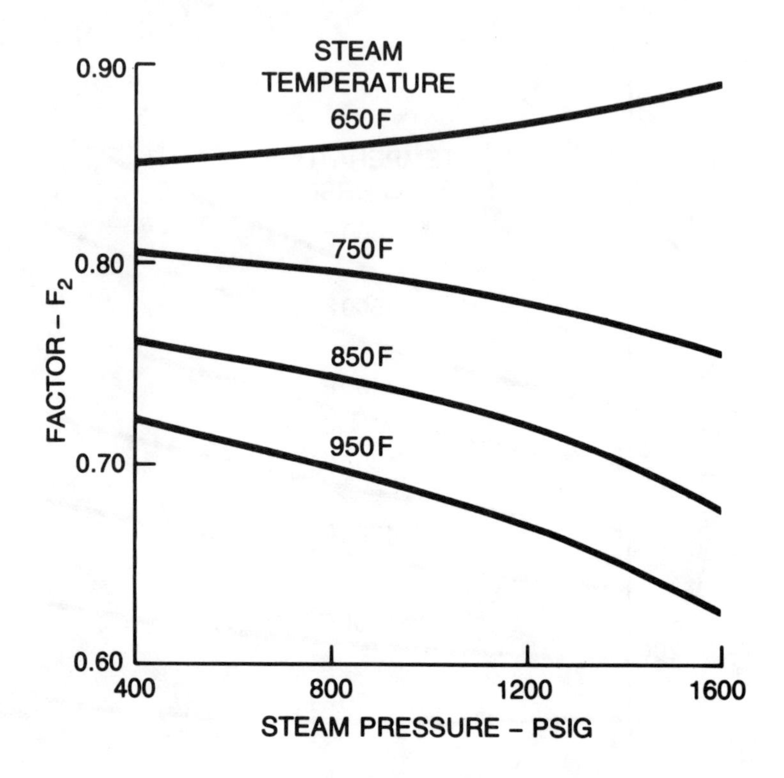


**Figure 8-11. Superheat Adjustment Factor**

*Cycle Design Flexibility*

One method of displaying the many options available using a gas turbine as a cogeneration application is shown in Figure 8-13. This diagram has been developed for the General Electric Company MS7001E gas turbine-generator (78,400 kw ISO, natural gas fired). A summary of the performance used to develop the performance envelope given in Figure 8-13 is presented in Table 8-2.

Point A represents the gas turbine-generator exhausting into an unfired, low pressure HRSG. Point C is a combined cycle configuration based on use of a two-pressure level unfired HRSG. The steam turbine in the C cycle is a noncondensing unit expanding the HP HRSG steam to the 150 psig process steam header.

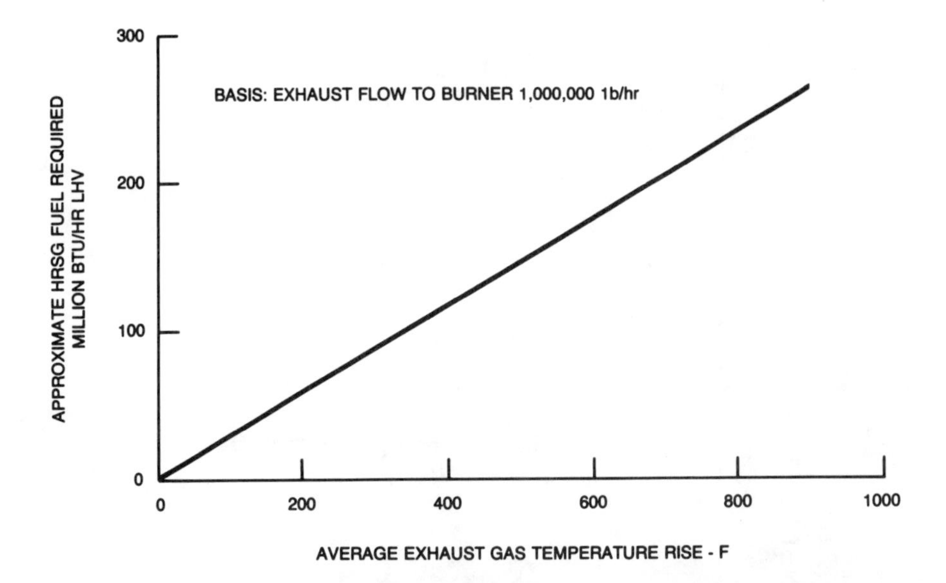


**Figure 8-12. Estimated HRSG Fuel Requirements**

Points B and D in Figure 8-13 represent operation of the HRSG with supplementary firing to a 1400F average exhaust gas temperature entering the heat transfer surface. The temperature used for the HRSG firing in Figure 8-13 has been arbitrarily limited to 1400F even though considerably higher firing temperatures and thus steam production rates are possible in the exhaust of this unit.

The "envelope" defined by A, B, C, D in Figure 8-13 represents most effective use of gas turbine in a cogeneration application. Operation along the line CE, DF, or any intermediate point to the left of line CD, represents use of condensing steam turbine power generation, with line EF applicable for combined cycle operation without any heat supplied to process. Thus, the cycles along line EF are combined cycles providing power alone.

The per unit cost of power generation for Cycles A through F illustrated in Figure 8-13 are presented in Figure 8-14. These per unit power generation costs represent the value required to generate a 25 percent discounted rate of return (DROR) on the investment for

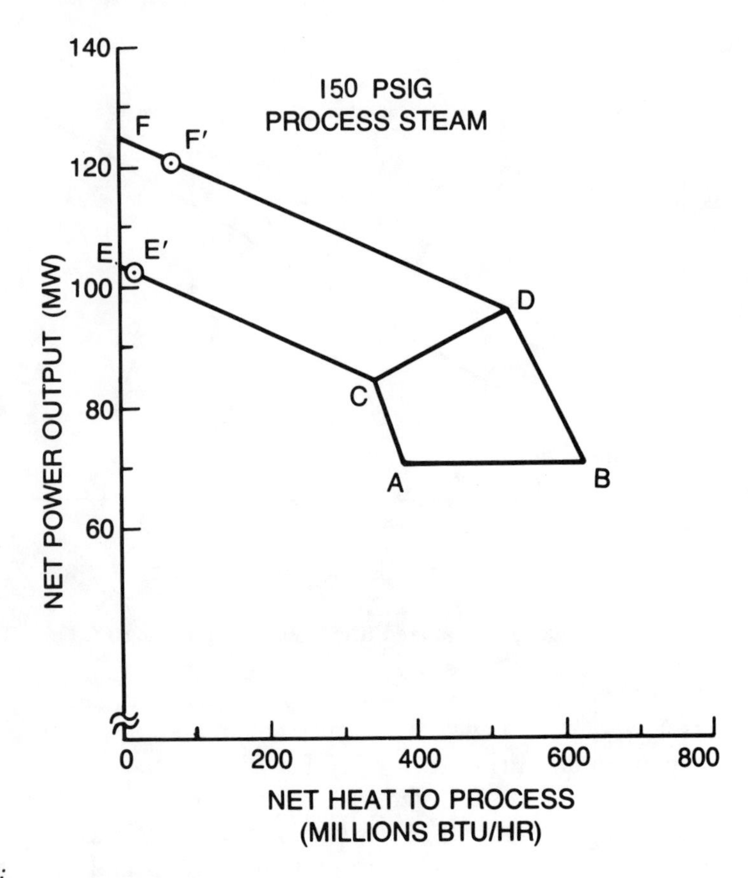


*Basis:*

1) Sea Level site, 80F ambient, natural gas fuel.
2) Cycle A—Unfired low pressure HRSG.
   Cycle B—Supplementary fired (1400F) HRSG, Low pressure process steam.
   Cycle C—Combined cycle, unfired, 2-pressure level HRSG, HP at 1450 psig, 950F, LP at 150 psig saturated, noncondensing steam turbine-generator.
   Cycle D—Combined cycle, supplementary fired HRSG, steam at 1450 psig, 950F, noncondensing steam turbine-generator.
   Cycle E—Same as Cycle C but with extraction/admission condensing steam turbine-generator.
   Cycle F—Same as Cycle D except with straight condensing steam turbine-generator.
3) Process returns and makeup enter the 5 psig deaerating heater at a mixed temperature of 180F.
4) Cycles E′ and F′ represent minimum extraction for process at 150 psig to meet requirement of a qualifying cogeneration facility under PURPA.

**Figure 8-13. Performance Envelope for Various Gas Turbine Cogeneration Systems MS7001E Gas Turbine-Generator**

**Table 8-2. Performance of Gas Turbine Cogeneration Cycles—MS7001E**

| *Cycle* | *A* | *B* | *C* | *D* | *E* | *F* |
|---|---|---|---|---|---|---|
| Net Output—MW | 70.4 | 70.1 | 84.2 | 96.4 | 103.1 | 124.4 |
| NHP—MBtu/hr | 382 | 627 | 346 | 521 | 0 | 0 |
| Net Fuel—MBtu/hr HHV | 396 | 370 | 439 | 496 | 851 | 1117 |
| FCP—Btu/kwh HHV | 5620 | 5280 | 5210 | 5150 | 8250 | 8980 |
| Power per Unit Heat to Process—kw/MBtu/hr | 184 | 112 | 243 | 185 | NA | NA |

*Basis:*

1) Cycle definitions as given in Figure 8-13.
2) Net output is the total power credited to the cogeneration cycle.
3) Net fuel includes credit for the net heat to process (NHP) at an 84% process boiler efficiency.

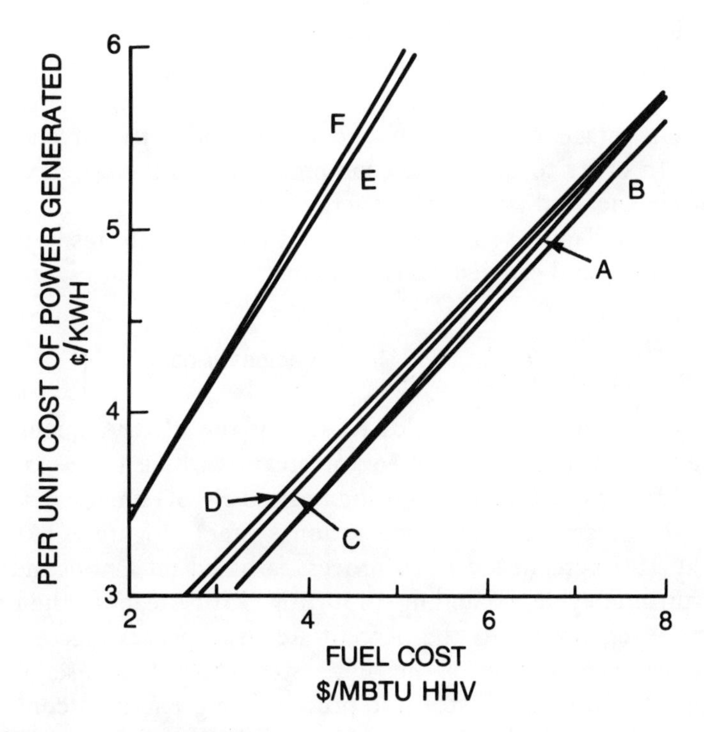


*Basis:*

1) Add-on to an existing facility.
2) Cycles as defined in Figure 8-13.
3) Operation 8400 hr/yr.

**Figure 8-14. Per Unit Cost of Power Generation MS7001E Cogeneration Systems—25% Discounted Rate of Return**

the cycle presented. Stated another way, applications where power has a value equal to or greater than that shown in Figure 8-14 will yield a DROR of 25 percent or more for the specific cycle configuration being examined.

The "pure" cogeneration cycles (cases A through D)—cycles without any condensing steam turbine power generation—have a much lower per unit cost threshold than the combined cycle designed to produce only power (cases E and F). Even so, site-specific fuel and power cost considerations may dictate cycles with considerable condensing power as the appropriate economic choice.

## Effects of PURPA

The legislation having the greatest impact on the development of cogeneration cycles is the Public Utility Regulatory Policies Act (PURPA), discussed in Chapter 1. The broad objectives of PURPA are to encourage conservation and the effective use of energy resources. In order to derive the economic benefits associated with the sale of cogenerated power, the energy supply system configurations being evaluated must satisfy basic criteria identified in this legislation. These thermal and efficiency criteria are summarized in Figure 8-15.

### *Cogeneration Cycles and PURPA Efficiency Requirements*

The ability of various cycles to satisfy the PURPA qualification criteria is shown in Figure 8-16 for steam turbine cycles, and Figure 8-17 for gas turbine cycles using unfired HRSG units.

For the steam turbine cogeneration system (Figure 8-16), operation at 100 percent steam to process is based on a noncondensing steam turbine cycle expanding 1450 psig, 950F steam to the process pressure levels noted. As the percent steam to process decreases, the cycle expands a portion of the boiler steam to a condenser with the limiting condition of no steam to process being a straight condensing steam turbine cycle based on 1450 psig, 950F initial steam conditions. In order to meet the qualification criteria of PURPA, this steam cycle would have to deliver at least 50 percent steam for a 50 psig process pressure level or 57 percent steam if the plant steam

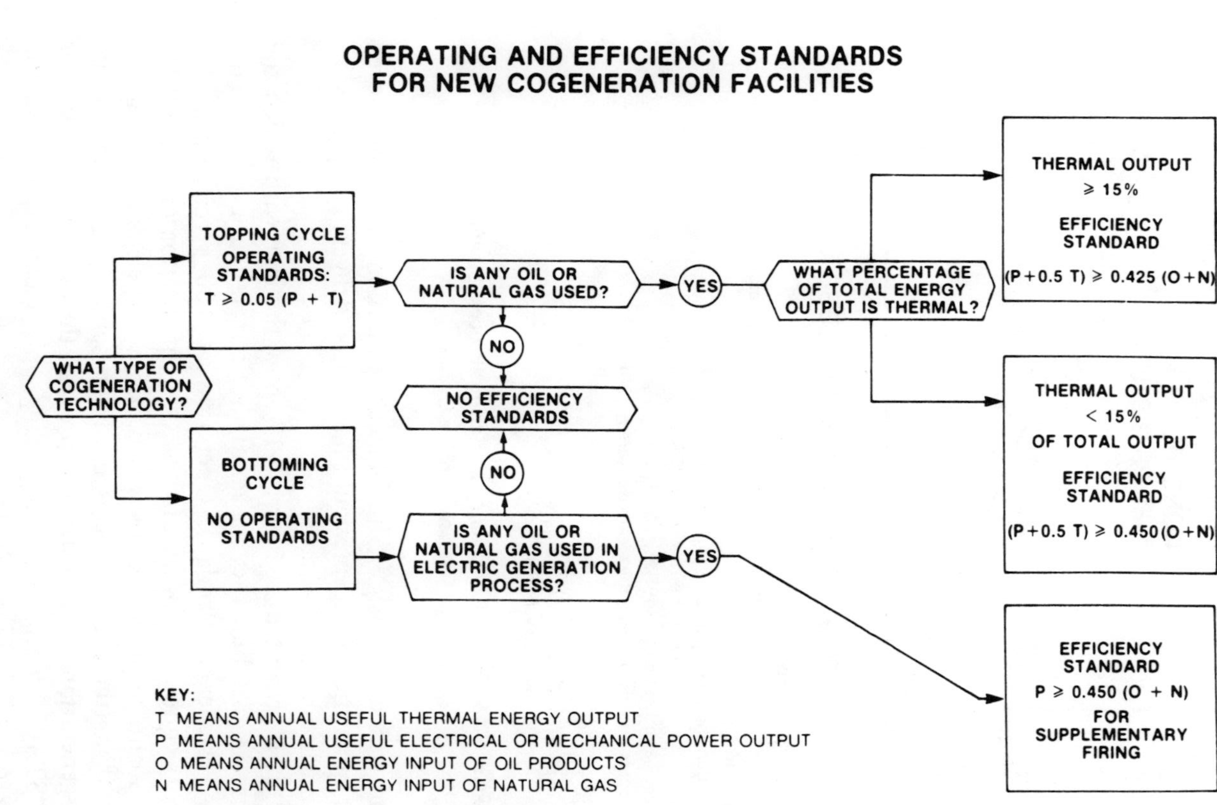


**Figure 8-15. Operating and Efficiency Standards for New Cogeneration Facilities**

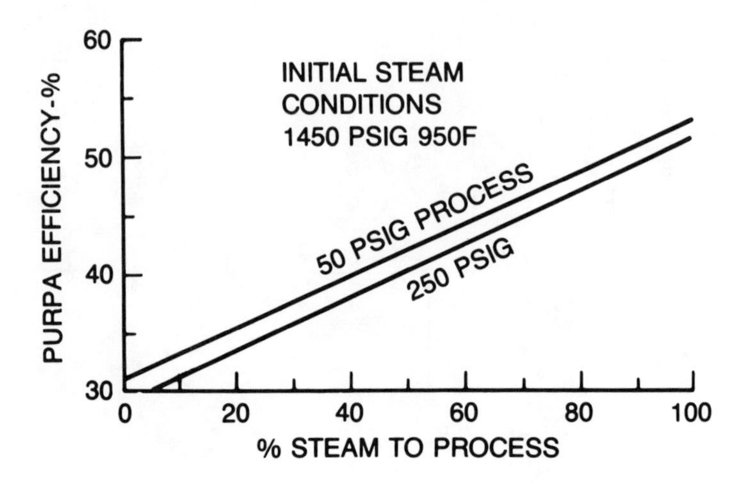


*Basis:*

1) Use of extraction condensing steam turbine-generator expanding steam to process pressures noted.
2) Condensing pressure is 2½" HgA.
3) Steam turbine-generator efficiency is 75%.
4) Feedwater heating to 445F.
5) Process returns 100% of the steam delivered, 180F.

6) $\text{PURPA } \eta = \dfrac{\text{Power} + \frac{1}{2}\text{ NHP}}{\text{Fuel (LHV Basis)}} \ (100)$

**Figure 8-16. PURPA Efficiency at Various Process Steam Demands—Steam Turbine Cycles**

demand was at 250 psig. If initial steam conditions were lower, even greater quantities of steam would have to be delivered to process to attain qualifying facility status.

The data presented in Figure 8-17 show the ease with which the gas turbine combined cycle configurations noted can meet the PURPA efficiency standard. Operation at 100 percent steam to process is equivalent to use of the "C" configuration in Figure 8-13. As the steam to process is decreased, more of the exhaust energy recovery from the gas turbine is used for condensing steam turbine power generation. At zero flow to process, the cycle operates similar to a "STAG" configuration providing only electric power to the system (point E in Figure 8-13).

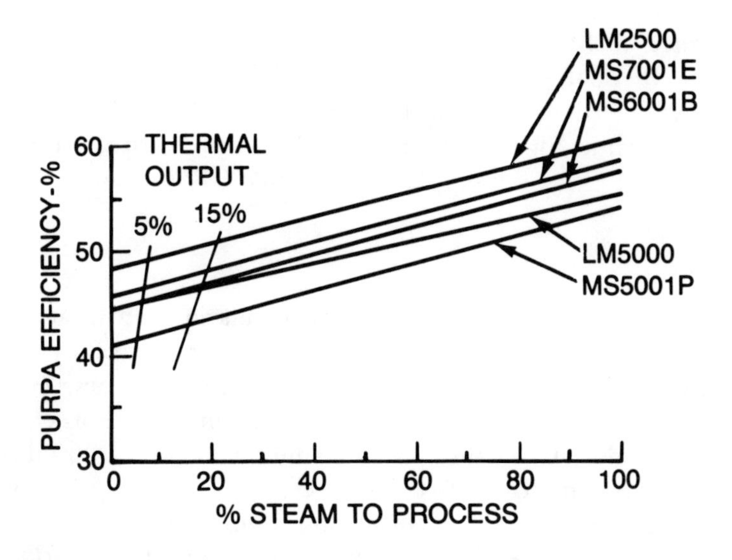


*Basis:*

1) Basic gas turbine information given in Table 8-2.
2) Sea level site, operation at 80F ambient capability, natural gas fuel.
3) Based on 2-pressure level unfired HRSG. HP HRSG is 850 psig, 825F for all cases except the LM5000 which is developed at 600 psig, 750F. LP HRSG feeds 150 psig steam header.
4) Process steam demand at 150 psig, 100% returns at 180F.
6) Condenser pressure is 2½" HgA in applicable cases.
7) PURPA $\eta = \frac{\text{Power} + \frac{1}{2}\text{ NHP}}{\text{Fuel (LHV Basis)}}(100)$
8) Thermal output $= \frac{\text{NHP}}{\text{NHP} + \text{Power}}(100)$

**Figure 8-17. PURPA Efficiency at Various Process Steam Demands— Gas Turbine Combined Cycles**

Note that all configurations except the MS5001P meet the 45 percent PURPA efficiency while supplying only 5 percent thermal energy to the process, the minimum quantity allowed under PURPA. And the MS5001P cycle meets the 42.5 percent PURPA efficiency criteria at a 15 percent thermal energy output to process.

These data show that present legislation permits gas turbine cycles to be developed providing a small amount of steam to process, and still receive qualifying facility status under PURPA because of their favorable thermal performance. (The minimum heat to process to comply with PURPA qualification criteria for the MS7001E system envelope given in Figure 8-13 is noted E′ for the unfired HRSG configuration and F′ for the supplementary fired HRSG system.)

*Examples*

The favorable effect of the PURPA regulations on what utilities must pay for power from qualifying facilities can have a profound influence on the development of cogeneration facilities, even for applications having small process heating demands. For example, assume an industrial facility has a requirement for 45,000 lb/hr of 150 psig saturated steam. The data presented in Table 8-3 show the largest cogeneration system providing about 100 MW of electric power would yield the highest discounted rate of return (DROR) for the specific conditions given except at $8/MBtu HHV fuel with a 7¢/kwh credit for all power generated. The most efficient cycle, the LM2500, is economically preferred only at the high fuel cost case. However, the DROR is less than 15 percent and the project would probably not meet the minimum criteria for discretionary investments.

An example for a somewhat larger industrial facility is given in Table 8-4. In this instance, the process heat demand is an ideal match with the thermal energy available from the exhaust of the LM2500. Thus, it provides not only the best FCP but also the most favorable economics for scenarios where the fuel cost is equal to $5/MBtu HHV or higher when power has a value of 7¢/kwh.

If the plant power requirement for the Table 8-4 example was 60 MW, and plant management was not interested in generating "excess" power, the LM2500 cogeneration system would provide the most favorable economics. However, the MS6001B system would also be a potential candidate, and its DROR may be sufficiently attractive to qualify for management consideration if adequate funds are available for this larger cogeneration system.

The third example presented in Table 8-5 is based on an industrial heat demand that is a match for the MS6001B in a combined cycle

**Table 8-3. Cogeneration Example—45,000 lb/hr Process Steam Demand**

| *System* | *LM 2500* | *MS6001B* | *MS7001E* |
|---|---|---|---|
| Net Output–MW | 24.1 | 47.2 | 100.6 |
| Net Fuel–MBtu/hr HHV | 157.1 | 372.4 | 796.9 |
| FCP–Btu/kwh HHV | 6520 | 7890 | 7920 |
| Estimated Total Installed Cost–$ Millions | 15.5 | 22.3 | 35.9 |
| DROR–% | | | |
| @ $4/MBtu HHV & 7¢/kwh | 37.0 | 42.5 | 54.0 |
| @ $6/MBtu HHV & 7¢/kwh | 26.5 | 26.0 | 34.0 |
| @ $8/MBtu HHV & 7¢/kwh | 14.8 | 3.5 | 6.8 |

*Basis:*

1) Process steam demand at 150 psig saturated.
2) Natural gas fuel, sea level site, 80F ambient.
3) 1983 investment costs for adding the cogeneration system to an existing industrial boiler plant with low pressure boilers. Costs include electrical substation and transformation to 138 kv.
4) Maintenance costs are 2.5% of the estimated total installed cost. Operating labor is $200,000 more than the labor cost without cogeneration.
5) Makeup water cost is $2/1000 gal. based on 50% loss of steam delivered to process for combined cycles. Makeup for cooling tower system is 5¢/1000 gal.
6) Operation 8400 hr/yr.
7) DROR based on 10% investment tax credit, 5-year depreciation, straight line method, 20-year economic life, no salvage value, 3% total property taxes and insurance, 50% income tax rate. All comparisons are with case without cogeneration.
8) Net output includes credit for auxiliary power requirements displaced due to the addition of the cogeneration system.
9) Net fuel includes a credit for fuel required in 84% efficient process boilers providing steam to process at 150 psig sat.
10) All cycles meet "qualification facility" requirements under PURPA.
11) Cycles

| | *LM2500* | *MS6001B* | *MS7001E* |
|---|---|---|---|
| HRSG (unfired) | | | |
| HP (psig/F) | 850/825 | 850/825 | 1450/950 |
| LP (psig/F) | 150/sat. | 150/sat. | 150/sat. |
| Steam turbine | | | |
| Extraction Condensing at 3" HgA | ~5 MW | ~14 MW | ~30 MW |

**Table 8-4. Cogeneration Example—85,000 lb/hr Process Steam Demand**

| *System* | *LM2500* | *MS60018* | *MS7001E* |
|---|---|---|---|
| Net Output—MW | 19.5 | 44.1 | 98.4 |
| Net Fuel—MBtu/hr HHV | 109.5 | 301.6 | 749.3 |
| FCP—Btu/kwh HHV | 5615 | 6840 | 7615 |
| Estimated Total Installed Cost—$ Millions | 9.4 | 21.0 | 35.4 |
| DROR—% | | | |
| @ $4/MBtu HHV & 7¢/kwh | 52.5 | 46.2 | 55.0 |
| @ $6/MBtu HHV & 7¢/kwh | 42.0 | 32.8 | 35.5 |
| @ $8/MBtu HHV & 7¢/kwh | 31.0 | 16.5 | 11.8 |

*Basis:*

1) See items 1 through 10 of Table 8-3.
2) Cycles

| | *LM2500* | *MS6001B* | *MS7001E* |
|---|---|---|---|
| HRSG (unfired) | | | |
| HP (psig/F) | NA | 850/825 | 1450/950 |
| LP (psig/F) | 150/sat. | 150/sat. | 150/sat. |
| Steam Turbine | | | |
| Extraction Condensing at 3" HgA | NA | ~11 MW | ~28 MW |

configuration. Even so, economics favor the larger MS7001E at fuel costs to about $6.15/MBtu HHV. Once fuel costs exceed this value, the MS6001B cogeneration cycle would be economically preferred.

All examples are based on adding a cogeneration system to an existing process plant. Thus, there was no "offsetting" process boiler plant investment that could be credited to the cogeneration plant investment as is the case when cogeneration is considered in new or grassroots facilities. If the evaluations were based on new facilities, the investment credit for the process boiler plant would have the most favorable effect on the gas turbine alternatives most consistent with the plant heat demands.

For example, for the Table 8-5 plant situation, the offsetting process boiler plant investment would reduce the incremental cogeneration investment for the LM2500, MS6001B and MS7001E alternatives to 68 percent, 75 percent and 86 percent, respectively, of the values

**Table 8-5. Cogeneration Example—180,000 lb/hr Process Steam Demand**

| *System* | *LM2500* | *MS6001B* | *MS7001E* |
|---|---|---|---|
| Net Output—MW | 26.6 | 39.6 | 92.9 |
| Net Fuel—MBtu/hr HHV | 129.3 | 208.1 | 632.6 |
| FCP—Btu/kwh HHV | 4860 | 5255 | 6810 |
| Estimated Total Installed Cost—$ Millions | 15.4 | 19.3 | 33.9 |
| DROR—% | | | |
| @ $4/MBtu HHV & 7¢/kwh | 43.0 | 50.0 | 59.0 |
| @ $6/MBtu HHV & 7¢/kwh | 35.5 | 40.7 | 41.6 |
| @ $8/MBtu HHV & 7¢/kwh | 27.3 | 30.0 | 22.3 |

*Basis:*

1) See items 1 through 10 of Table 8-3.
2) Cycles

| | *LM2500* | *MS6001B* | *MS7001E* |
|---|---|---|---|
| HRSG | Fired | Unfired | Unfired |
| HP (psig/F) | 850/825 | 850/825 | 1450/950 |
| LP (psig/F) | NA | 150/sat. | 150/sat. |
| Steam Turbine Type | Noncond ~7 MW | Noncond ~6 MW | Extr Cond ~22.5 MW |

given in Table 8-5. The magnitude of the reduction in investment would have a more favorable effect on the LM2500 and MS6001B alternatives than for the MS7001E cogeneration system.

## Conclusion

This chapter has briefly reviewed technical considerations in the application of steam turbine and/or gas turbine cogeneration systems. These basic principles combined with sound cost estimating procedures can provide system designers an effective means of quickly defining which alternatives merit serious consideration.

Years ago, the appropriate cogeneration system for a specific application was significantly influenced by its "fit" into the plant heat and power requirements. However, the promulgation of PURPA introduces an additional "degree of freedom" since cycle power generating capability relative to the heat that can be supplied to process is no longer significantly constrained. Furthermore, economic analyses are illustrating that in many applications this is the appropriate method of cogeneration cycle development.

# Chapter 9

## Flexibility and Economics of Combustion Turbine–Based Cogeneration Systems

***M. V. Wohlschlegel, G. Myers and A. Marcellino***

Today, there has been a resurgence of interest in cogeneration systems. This resurgence has been driven by a number of factors. The legislative influence with the enactment of the National Energy Act of 1978 has provided incentives for industry by removing regulatory and institutional obstacles to cogeneration and providing economic incentives for its implementation. From a strategic perspective, many industries have assessed the potential competitive advantages associated with cogeneration. In particular, industries which are highly energy intensive, i.e. a large part of the value-added of their product is electrical and/or thermal energy, have recognized the strategic implications of cogeneration systems in their businesses and long-term profitability.

This chapter will demonstrate how a combustion turbine-based cogeneration system can be incorporated into a representative industrial process. The economic and operational benefits associated with the cogeneration system are discussed. Options to the basic system which further integrate combustion turbines into the process are presented along with the resultant total and incremental benefits of each.

**Technical Evaluation Parameters**

Two parameters which must be defined when comparing various heat and power cycles are efficiency and heat rate. These terms are typically employed with little confusion when evaluating a cycle or system which singularly produces either thermal energy or electrical energy. For systems which coincidentally produce both thermal and electrical power, a new definition must be considered.

*Efficiency*

A general definition for system efficiency is simply useful energy or work output divided by total energy input.

For purposes of cogeneration systems, useful energy will be considered to include both net electrical energy (gross electrical energy less auxiliaries) and net thermal energy transferred to an industrial process. Inefficiencies and losses within the industrial process will be ignored since they are not affected by the energy source.

The efficiency *(E)* is expressed as follows:

$$(E) = \frac{\text{Net Electrical Energy} + \text{Useful Thermal Energy}}{\text{Fuel Energy}} \times 100$$

or

$$(E) = \frac{\begin{matrix}\text{Net Pwr} \\ \text{(kw)}\end{matrix} \times 3412 \left(\frac{\text{Btu}}{\text{kw-hr}}\right) + \text{Useful Thermal Energy} \left(\frac{\text{Btu}}{\text{hr}}\right)}{\text{Total Fuel HHV Input} \left(\frac{\text{Btu}}{\text{hr}}\right)} \times 100$$

*Cogeneration Heat Rate*

Heat rate has traditionally been defined for conventional electric power plants as thermal energy input divided by electrical output. When defining Cogeneration Heat Rate (HRcg), the additional factor to be considered is the displaced fuel which otherwise would have been consumed in a conventional process boiler to provide the required process thermal energy.

The displaced fuel energy is the total process thermal energy demand divided by the process boiler efficiency. Thus, the formula for Cogeneration Heat Rate is:

$$HRcg = \frac{\text{Total Fuel Energy (HHV)} - \dfrac{\text{Process Heat}}{\text{Process Boiler Efficiency}}}{kw_{net}}$$

**Qualitative Value of Efficiency and Cogeneration Heat Rate**

Efficiency alone is not an adequate evaluation parameter when considering various cogeneration alternatives because it does not account for the relative economic worth of thermal and electrical energy. The economic value of electrical energy expressed in a thermal manner ($.06/kwh–$17.58 per million Btu) is more than three times that of thermal energy generated by burning fuel ($5.00/million Btu).

Cogeneration Heat Rate overcomes this limitation by factoring the efficiency at which the cogeneration plant uses additional fuel, above that originally required by the conventional boiler to meet the process heat needs, to produce additional electrical energy. Since fuel cost is the major factor in the overall plant operating cost equation, Cogeneration Heat Rate, when multiplied by the fuel cost results in an estimate of direct generating cost in dollars per kilowatt-hour. This cost can be contrasted to the industrial cogenerator's purchased power cost to obtain an indication of the economic viability of the cogeneration project.

**Economic Evaluation Techniques**

Financial analysis methodologies employed to evaluate cogeneration projects are varied. In the following report, a mathematical model based on the discounted cash flow (DCF) and the more commonly used simple payback evaluation techniques are employed.

For a given cogeneration cycle configuration, the annual cash flows are discounted and summed over the assumed economic life of the project to arrive at the Net Present Value (NPV). The higher the

NPV, the higher is the project's contribution to the net worth of the equity investor.

The internal rate of return (IRR) is also calculated. IRR is the rate that equates the present value of the cash outflows and inflows, or the rate of return on invested capital which the project is returning to the firm given that the net cash inflows each year are reinvested at the IRR.

If the firm cannot reinvest the cash inflows at the IRR rate, the IRR value overstates the true rate of return on the project. The simple payback is calculated for each cogeneration cycle configuration. Caution must be exercised when using the payback technique since this method does not consider all cash flows nor does it discount them.

The variable inputs to the model include initial investment outlay, expected revenues and expenses over the economic life of the project, and the associated effects of tax and depreciation on the annualized cash flows. The discounting factor employed is an after-tax weighted average cost of capital which is derived from variable leverage inputs. For purposes of this evaluation, a leverage ratio of 70 percent of total investment has been assumed.

The model employed serves its purpose for the analysis; however, care should be taken in its universal employment. The model as displayed is not suitable for rigorous financial decision-making within the industrial capital budgeting process due to the following limitations:

- The after-tax weighted average cost of capital discounting factor comingles the investment and financing decision processes.
- It assumes a single cost of capital throughout the economic life of the project which implicitly assumes a constant cost of capital, a constant reinvestment rate, and an ever-increasing financial risk over the life of the project.

The model provides variable inputs for such items as boiler fuel cost, cogeneration prime mover fuel cost, operation and maintenance costs, and purchased power costs, with an assumed escalation value for each over the project's economic lifetime.

The key assumptions employed to evaluate each cycle configuration are shown in Table 9-1.

**Table 9-1. Key Assumptions**

| | |
|---|---|
| Economic Life | 10 years |
| Debt | 70% |
| Debt Cost | 14% |
| Return on Equity | 18% |
| Weighted Average After Tax Cost of Capital | 10.49% |
| Property Tax Rate | 1.5% |
| Insurance Rate | 1.5% |
| Investment Tax Credit | 10% |
| Income Tax Rate | 48% |
| Tax Depreciation | 5 years |
| Conventional Boiler Fuel Cost | $3.60/MBtu (HHV) |
| Cogeneration Fuel Cost | $4.00/MBtu (HHV) |
| O&M Cost | $2.50/MWH |
| Cost of Purchased Electricity | $65.00/MWH |
| Price of Sold Electricity | $55.00/MWH |
| Conventional Boiler Efficiency | 84% |
| Capacity Factor | 92% |

## Typical Process Suitable for Cogeneration

A simplified but representative industrial process providing steam for process heat is shown in Figure 9-1. The conventional boiler generates 725,000 pounds per hour (pph) of 600 psig 750°F steam. The boiler is fueled by distillate oil which is burned at the rate of 1020 million Btu/hr at an efficiency of 84%. Thermal energy is recovered (856 million Btu/hr) from the steam at three pressure levels (600 psig, 300 psig, 5 psig) by two process heat exchangers, a mechanical drive backpressure steam turbine, and a deaerating feedwater heater. All of the steam is generated at the highest pressure level required. Steam used at lower pressure is supplied either through pressure reducing stations or the backpressure steam turbine. In addition to these steam requirements, the process under discussion has an assumed average demand of 45,000 kw which is supplied by the serving electric utility company.

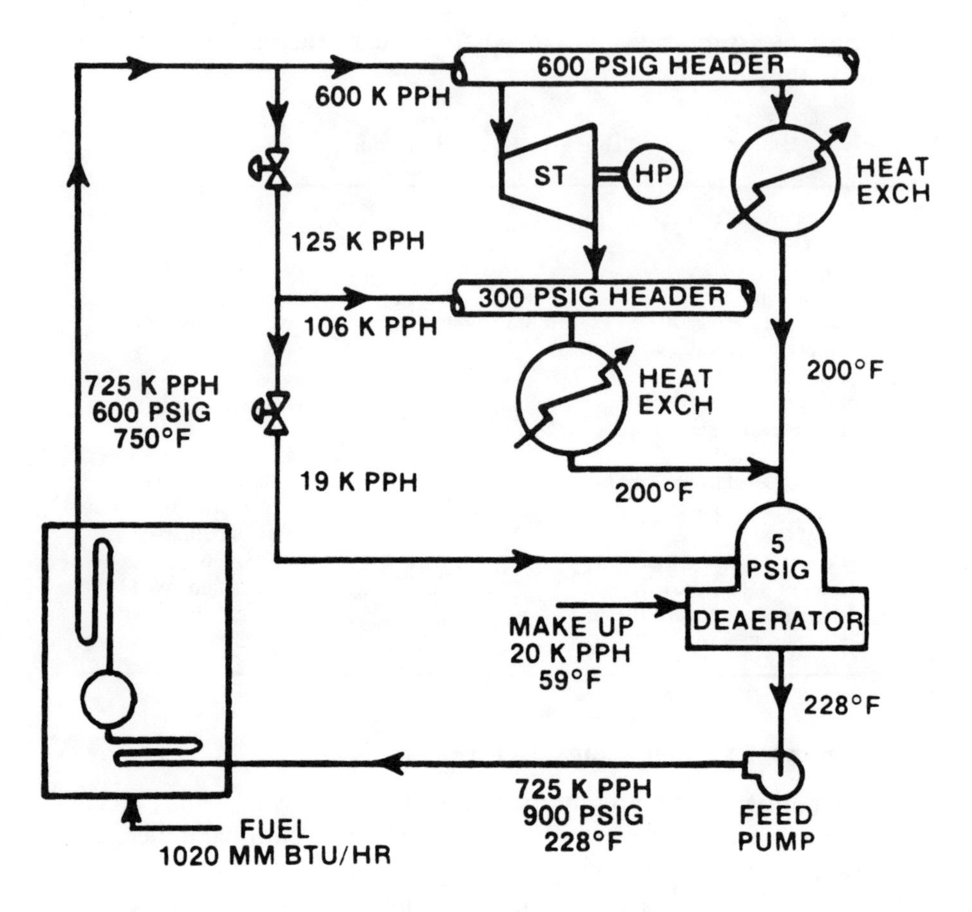


Figure 9-1. Typical Process Plant

## Combustion Turbine Cogeneration—Equipment Requirements

The major items of equipment required to incorporate a combustion turbine cogeneration system into an industrial process plant are discussed below.

### *Combustion Turbine Package*

For the example, a Westinghouse Model W501D industrial heavy-duty combustion turbine generator package is employed. Typically, this system would include the following:

- W501D turbine with an ISO baseload rating including inlet and exhaust losses of 99,750 kw at the generator terminals and a high heating value heat rate of 11,660 Btu/kwh when burning natural gas fuel
- Hydrogen-cooled generator with brushless excitation system
- Electric motor starting package
- Inlet air system including ductwork and silencing
- Electrical/control auxiliary package including control system, motor control centers, protective relay panels, etc.
- Mechanical package including fuel and lube oil systems
- Coolers for generator hydrogen, lube oil and turbine cooling air
- Pipe racks between auxiliary packages and turbine and generator
- Enclosures for turbine and auxiliary packages
- Instrumentation for control and protection of the turbine, generator and auxiliaries.

*Heat Recovery Steam Generator Package*

The Heat Recovery Steam Generator (HRSG) is uniquely designed to meet the specific needs of the industrial process. The HRSG package includes the following equipment:

- Heat Recovery Steam Generator with superheater, natural circulation evaporator, economizer, steam drum and interconnecting piping
- Ductwork from combustion turbine exhaust flange to HRSG inlet
- Exhaust stack
- Feedwater inlet and steam discharge piping with feedwater control valve, steam check and stop valves
- Electronic control system and all ASME code required instrumentation.

*Balance of Plant Equipment*

The scope of the balance of plant (BOP) equipment can vary significantly on cogeneration projects. The scope depends on many factors including the suitability of existing equipment and the users' preferences. Consequently, BOP costs can range from as little as 10% to as much as 50% of the total project cost. The equipment list shown below has been assumed in the scope for purposes of the cogeneration cycle evaluation performed.

- Equipment foundations (concrete, rebar, anchor bolts, etc.)
- Steam and feedwater piping systems from existing facilities to HRSG
- Conduit and wiring from cogeneration plant to existing facilities
- Service piping from HRSG to existing systems (chemical feed, blowdown, drains, instrument air, etc.)
- Gas fuel piping from combustion turbine to existing main
- Electrical equipment interconnects (transformers, switchgear, etc.)
- Central control room equipment
- Switchyard
- Outdoor lighting, communications, fences, fuel supply, fire protection.

The standard industrial combustion turbine package includes much of the BOP equipment; for example, the auxiliary power and compressed air requirements are satisfied internally and cooling water is not required. In addition, the plant systems supporting the existing boiler, such as water treatment, deaeration, blowdown, etc., can usually be diverted to satisfy the HRSG requirements.

*Engineering and Construction*

Combustion turbine cogeneration plants are, for the most part, composed of standard designed equipment which make use of modularization and shop fabrication to the greatest extent possible. Hence, engineering and erection costs are substantially lower than conventional electrical power and steam generating equipment of comparable sizes. The primary erection tasks would generally be comprised of the following:

- Site preparation (grading, pouring foundations, digging drainage ditches and cable trenches, etc.)
- Installation of combustion turbine package
- Installation of HRSG package
- Installation of BOP equipment including tie-ins to existing facilities.

*Total Cost of Cogeneration Plant*

The total cost of the cogeneration plant has been estimated to be $49.9 million. This includes the cost of all equipment, engineering, erection, interconnects with and modifications to existing facilities, as well as permits and licensing, and interest during construction. Analysis of the individual cost components reveal the following breakdown:

|  |  |
|---|---|
| • Major Equipment Packages (Combustion Turbine & HRSG) | 40% |
| • Balance of Plant Equipment | 25% |
| • Engineering and Construction | 15% |
| • Other (Interest during Construction, Permitting, Contingency, etc.) | 20% |
| | 100% |

**Cogeneration with Basic CT/HRSG Cycle**

In Configuration A (see Figure 9-2), the existing steam generation is supplemented by adding a combustion turbine and an associated heat recovery steam generator system. Note that in this configuration, the method of responding to fluctuations in process steam demand remains unchanged; i.e., by modulating the firing rate of the existing boiler. This increases the reliability of the steam supply since the loss

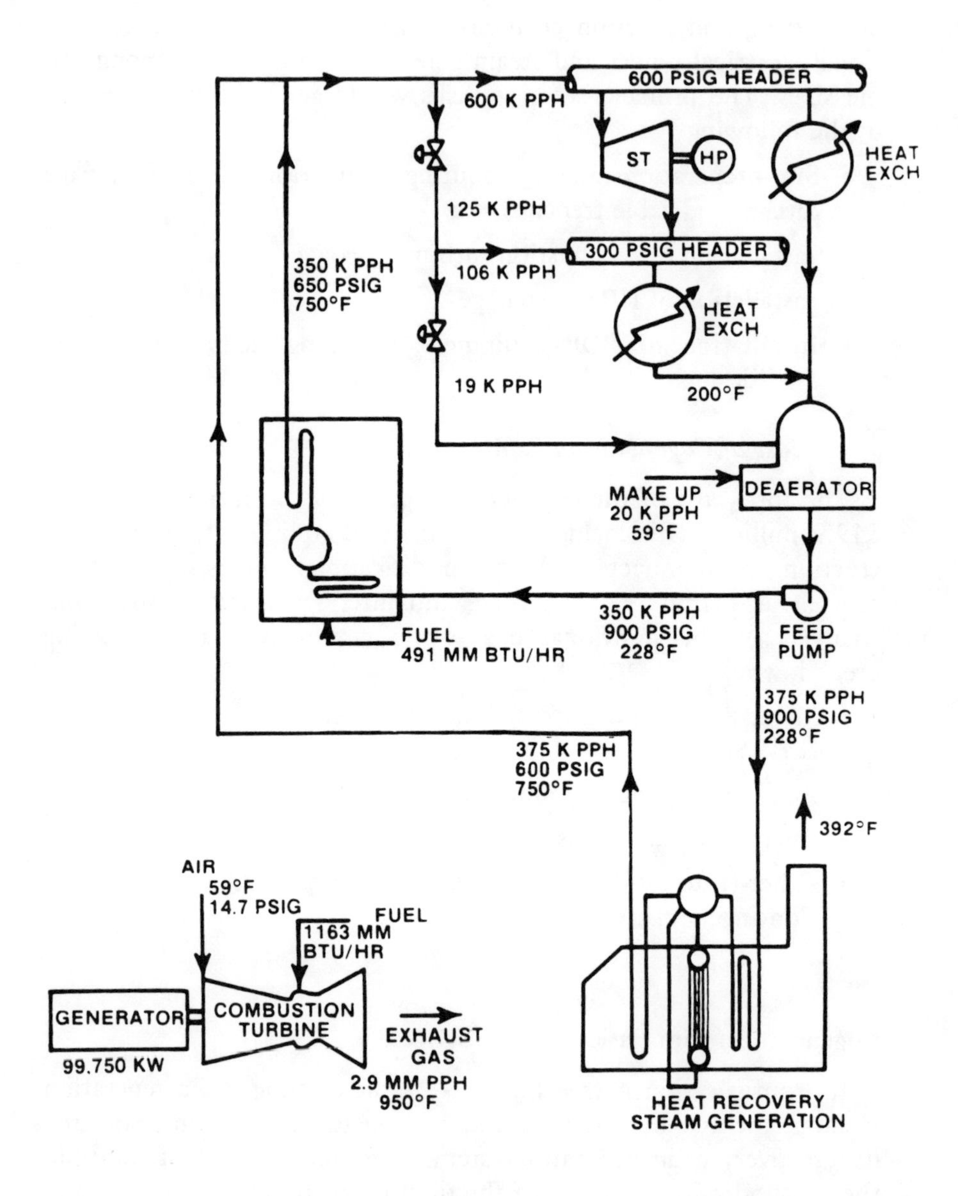


Figure 9-2. Configuration A—Cogeneration with Basic CT/HRSG Cycle

of either the HRSG or the conventional boiler does not result in the total loss of steam flow.

The thermodynamic cycle data and financial performance for Configuration A are summarized in Tables 9-2 and 9-3. The plant generates 99,750 kw of electrical power and 375,000 pph of 600 psig 750°F steam. The electrical power generated meets the 45,000 kw process requirements and provides a surplus of 54,750 kw for sale to the electric utility. The steam produced by the HRSG provides 52 percent of the process steam requirements with a corresponding reduction of fuel input to the existing boiler.

Although the total fuel consumption has been increased by 62 percent (1,654 million Btu/hr) and the overall thermal efficiency reduced from 84 percent to 72 percent, the additional fuel has been used to generate power at a cogeneration heat rate of 6356 Btu/kwh. The power produced displaces that previously purchased and provides an excess for sale to the utility, thereby generating revenue to offset the incremental fuel costs as well as other costs. As seen in Table 9-3, the $49.9 million-dollar investment in this basic CT/HRSG cogeneration cycle yields a net present value of $107.62 million, an IRR of 47.07% and a simple payback of 2.11 years.

## Cogeneration–Steam Production Increase by Ductburning

As evidenced in Configuration A, a single W501D combustion turbine generates more than double the electrical demand but only one-half of the steam requirements for the industrial process. This is typical of combustion turbine-based cogeneration systems because the high combustion turbine specific power (124 kw per lb/sec of exhaust gas for the W501D) results in a high power-to-steam ratio (266 kw per lb/hr of steam for the basic cogeneration plant–Configuration A).

In Configuration B (see Figure 9-3), the conventional boiler is entirely replaced by the combustion turbine/HRSG system. This configuration employs supplementary firing utilizing burners located in the HRSG inlet ductwork. This method of producing additional steam represents an extremely efficient utilization of incremental

**Table 9-2. Cogeneration Configuration Performance Summary**

| | *Power Generation and Use, kw* | | | | *Thermal Energy Generation and Use* | | | | |
|---|---|---|---|---|---|---|---|---|---|
| *Configuration* | *Combustion Turbine Net* | *Steam Turbine Net* | *Process Power Use* | *Power Sale* | *Heat to Process MMBtu/hr* | *Conventional Boiler Flow lbs/hr* | *HRSG Pressure psig* | *Temperature °F* | *Flow lbs/hr* |
| Base Process | 0 | 0 | 45,000 | 0 | 856 | 725,000 | 0 | 750 | 0 |
| A | 99,750 | 0 | 45,000 | 54,750 | 856 | 350,000 | 600 | 750 | 375,000 |
| B | 99,750 | 0 | 45,000 | 54,750 | 856 | 0 | 600 | 750 | 725,000 |
| C | 99,750 | 0 | 45,000 | 54,750 | 856 | 0 | 600 | 750 | 665,000 |
| | | | | | | | 300 | D/S | 71,000 |
| D | 99,750 | 11,450 | 45,000 | 66,200 | 856 | 0 | 1250 | 900 | 664,000 |
| | | | | | | | 300 | D/S | 73,000 |
| E | 111,250 | 0 | 45,000 | 66,250 | 856 | 0 | 600 | 750 | 657,000 |
| | | | | | | | 300 | D/S | 81,000 |

| | Fuel Use MMBtu/hr | | | | | Cycle Efficiencies | | |
|---|---|---|---|---|---|---|---|---|
| *Configuration* | *Boiler* | *Combustion Turbine* | *Duct Burner* | *Total* | *Increased Fuel Use* | *HRSG Efficiency (%)* | $\eta$ *Thermal Efficiency %* | *Cogeneration Heat Rate Btu/kwh* |
| Base Process | 1020 | 0 | 0 | 1020 | 0 | – | 84. | – |
| A | 491 | 1163 | 0 | 1654 | 634 | 77. | 72.3 | 6356 |
| B | 0 | 1163 | 417 | 1580 | 560 | 90. | 75.7 | 5614 |
| C | 0 | 1163 | 360 | 1523 | 503 | 96. | 78.6 | 5042 |
| D | 0 | 1163 | 406 | 1569 | 549 | 96. | 78.7 | 4937 |
| E | 0 | 1277 | 339 | 1616 | 596 | 94. | 76.5 | 5357 |

$$\eta = \frac{\text{Net kw} \times 3412 + \text{Process Heat}}{\text{Fuel Input Energy (HHV)}} \times 100$$

$$\text{Co-Gen. H.R.} = \frac{\text{Fuel Input Energy} \cdot \text{(Heat to Process/Conv. Boiler Efficiency)}}{\text{Net kw}}$$

$$\text{HRSG Efficiency} = \frac{\text{Gas Temp In} - \text{Gas Tem Out}}{\text{Gas Temp In} - \text{Feedwater Temp In}} \times 100$$

**Table 9-3. Economic Analysis Summary**

| *Analysis* | *Overall Analysis Summary* | | | | | *Incremental Investment Analysis Summary* | | | |
|---|---|---|---|---|---|---|---|---|---|
| *Cycle Configuration* | *A* | *B* | *C* | *D* | *E* | *B vs A* | *C vs B* | *D vs C* | *E vs C* |
| Capital Cost ($ × 1000) | 49,900 | 51,400 | 52,400 | 55,970 | 54,400 | 1,500 | 1,000 | 3,570 | 2,000 |
| Net Present Value ($ × 1000) | 107,624 | 118,511 | 126,168 | 142,533 | 136,126 | 10,887 | 7,657 | 16,365 | 9,958 |
| Simple Payback (yrs) | 2.11 | 2.09 | 2.06 | 2.02 | 2.03 | 1.27 | .80 | 1.22 | 1.23 |
| Internal Rate of Return (%) | 47.07 | 48.58 | 50.13 | 52.12 | 51.38 | 87.64 | 127.62 | 80.37 | 81.93 |

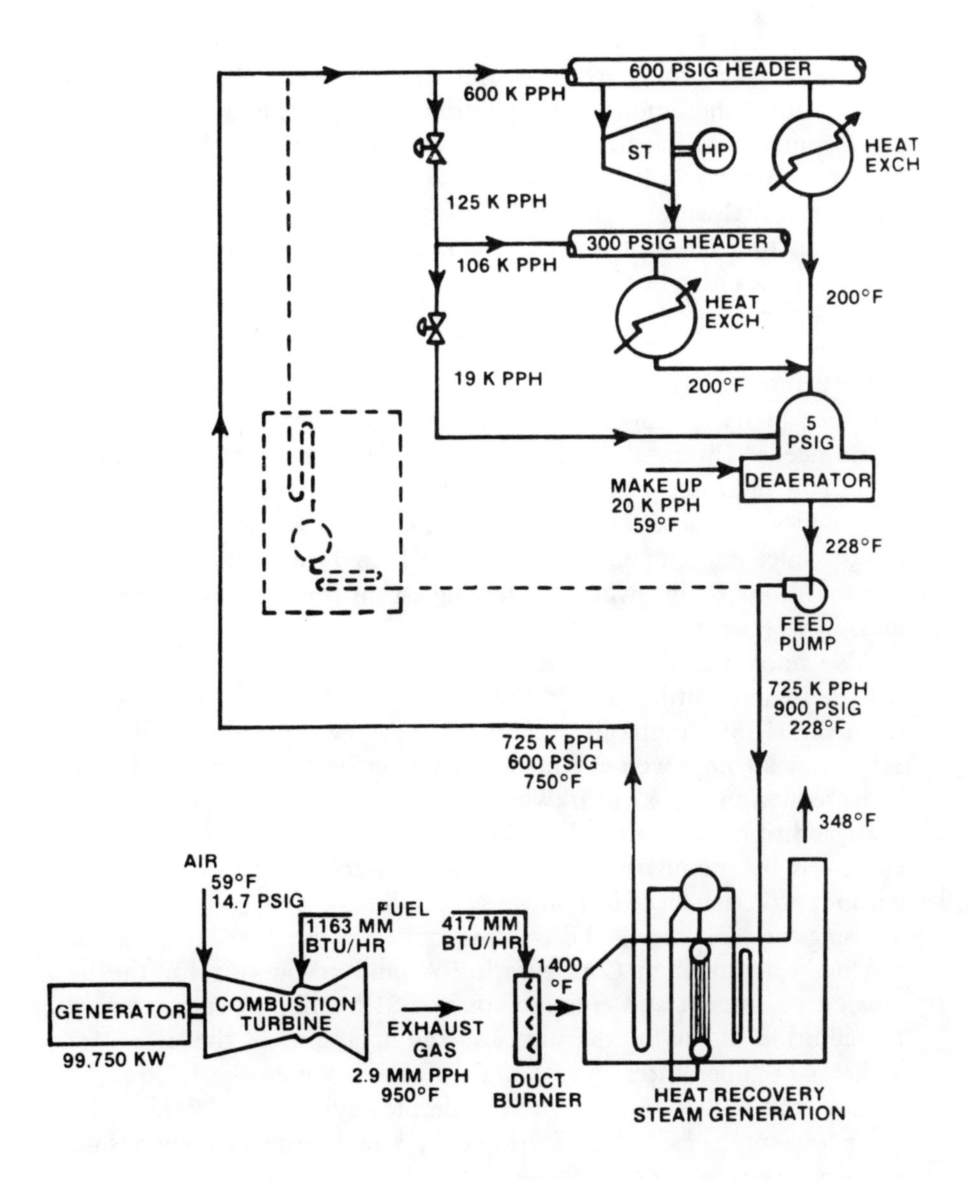


**Figure 9-3. Configuration B—Cogeneration Steam Production Increase by Duct Burning**

fuel energy—87% to 88% on any HHV basis. The only losses associated with ductburning are those due to moisture in the exhaust gas (11%) and conduction and radiation through the ductwork and steam piping. Sufficient oxygen is available in the combustion turbine exhaust gas (15%–16%) to allow combustion of a substantial amount of fuel without the need for supplemental air.

Hence, for the same stack temperature, the increased heat transferred to steam is not debited by increased stack loss as would be the case for a conventional boiler wherein increases in fuel input requires additional air to maintain combustion.

By firing the exhaust gas to 1,400°F, the HRSG steam production has been increased to 725,000 pph which satisfies all of the process steam needs. The resultant heat recovery efficiency has increased to 90%. This increase in heat recovery efficiency occurs because of the increased steam production in the HRSG and is achieved by increasing the inlet gas temperature from 950°F to 1,400°F as opposed to a conventional boiler where increasing steam flow requires increased mass flow.

Also note that the stack temperature decreased 44°F. The effective fuel energy utilization in the ductburner and lower stack losses from the HRSG results in an increase in overall thermal efficiency of 3.5% and an improvement in cogeneration heat rate to 5,614 Btu/kwh (reduction of 742 Btu/kwh or 11%).

In addition to improved cycle efficiency, operating flexibility is enhanced by modulating ductburner firing to match steam demand without affecting electrical power generation.

Using the same assumed economic factors as were used in the basic cogeneration plant in Configuration A, and accounting for the increased equipment and erection costs of $1.5 million associated with the addition of the ductburners and the upgrading of the HRSG for higher gas temperatures and steam flow, the NPV increases to $118.51 million, the IRR is 48.58%, and the simple payback is 2.09 years. On an incremental basis, the additional $1.5 million investment results in a simple payback of 1.27 years.

## Cogeneration Efficiency Increase— Dual Pressure HRSG

The discussion of cogeneration Configuration B revealed that, as a result of highly effective use of fuel in a ductburner and its effect on HRSG efficiency, replacing all of the conventionally generated steam with that produced in a supplementally fired HRSG improved the overall thermal efficiency of the cogeneration system. The increase in NPV signifies that the investment makes economic sense. These benefits can be extended by further increasing the HRSG heat recovery efficiency.

Although a thorough discussion of HRSG optimization is beyond the scope of this chapter, the improvement in heat recovery associated with a marginal increase in surface area cannot be economically justified. This is true because heat is absorbed in the HRSG solely by convection, the driving force being the temperature difference between the exhaust gas and the evaporator tubes which are essentially at the saturated steam temperature (and, hence, fixed by the boiler steam pressure).

This temperature difference is called the boiler "pinch point." As the boiler's heat transfer surface area is increased by adding successive rows of evaporator tubes, the temperature difference between the exhaust gas and the tubes is reduced. This results in progressively less heat transfer for each increment in surface area so that eventually, the exhaust gas temperature is so close to the saturation temperature that theoretically an infinite amount of surface area is required to extract additional heat.

However, if steam can be utilized at lower pressure levels, the case for most industrial processes, the HRSG heat recovery efficiency can be effectively increased by adding another pressure level to the HRSG. The addition of a multiple pressure level HRSG to the representative industrial process is shown in Configuration C (see Figure 9-4). In this case, the steam pressure and the saturation temperature have been lowered allowing the additional surface area of the lower pressure evaporator and economizer tubes to absorb heat at the "restored" exhaust gas-to-steam/water temperature difference.

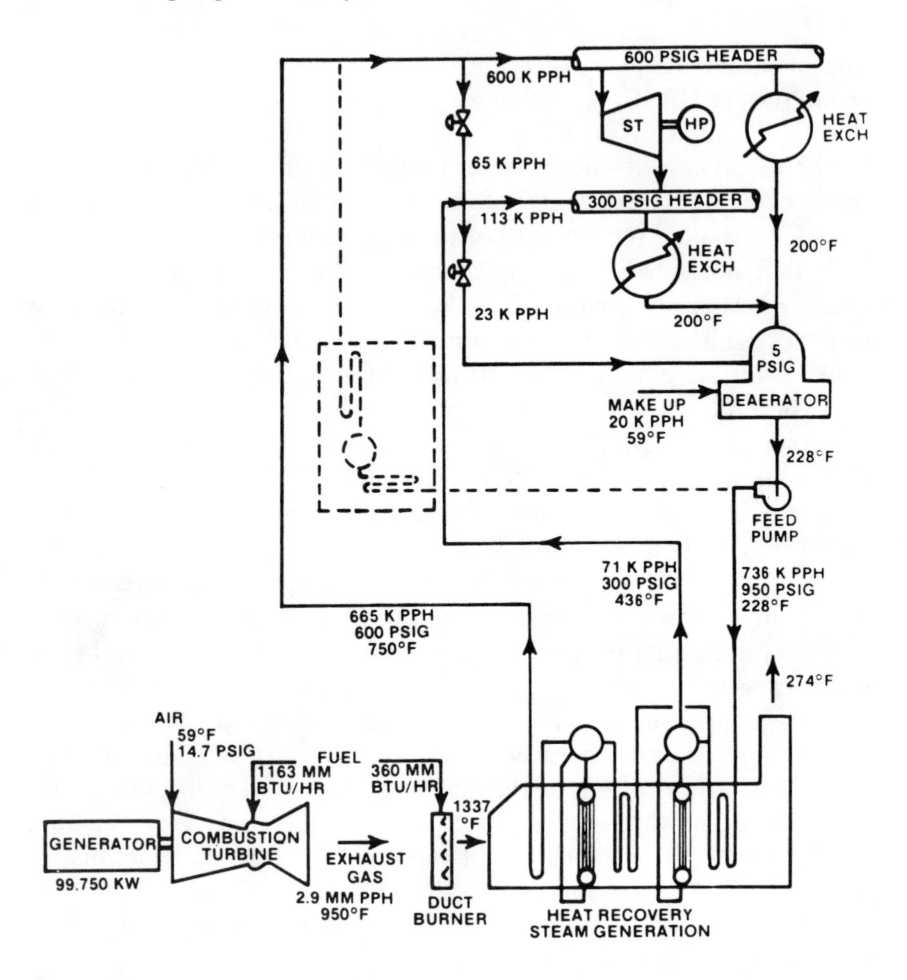


**Figure 9-4. Configuration C—Cogeneration Efficiency Increase by Dual Pressure HRSG**

The combined effects of inlet gas temperature and multiple pressure levels on the HRSG heat recovery efficiency are shown in Figure 9-5. This approach is not required in a conventional boiler due to its inherently high gas-to-steam/water temperature difference throughout the heat transfer surfaces.

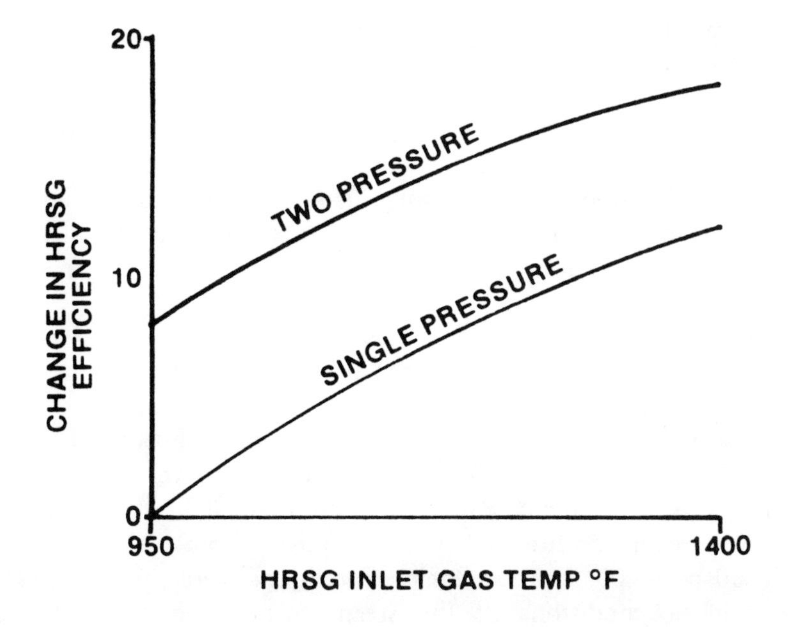


**Figure 9-5. HRSG Efficiency Variation**

Referring to Configuration C, the lower pressure level operates at 300 psig and generates 71,000 pph of steam satisfying roughly half of the steam required by the process at this pressure. This reduces the steam required at 600 psig so that the HRSG inlet gas temperature can be lowered to 1,337°F reducing the ductburner fuel by 15% (57 million Btu/hr).

As summarized in Table 9-2, the overall thermal efficiency has been increased by 2.9% and the cogeneration heat rate improved to 5,042 Btu/kwh (a reduction of 571 Btu/kwh or 10%). The additional equipment and erection cost associated with the addition of the added pressure level is estimated to be $1.0 million.

The net present value for the $52.4 million investment is $126.17 million with an IRR of 50.13% and a simple payback of 2.06 years. The incremental investment of $1.0 million results in an incremental simple payback of 0.8 years.

## Cogeneration Power Increase—Backpressure Steam Turbine

Thus far, methods of increasing the net present value through investments in cogeneration by improving the efficiency at which steam is generated (reducing fuel costs) have been discussed. The cogeneration decision can be made even more attractive by increased electrical generation which increases the revenues accrued from the sale of electricity.

An approach often taken in industrial processes is to add a backpressure steam turbine. This approach can also be incorporated into a Combustion Turbine-Based Cogeneration plant. This scheme was applied to the representative industrial process as shown in Configuration D (see Figure 9-6). The multiple pressure level HRSG design (introduced in Configuration C) has been retained; however, the high pressure steam conditions have been raised from 600 psig–750°F to 1250 psig–900°F. These steam conditions were selected so that the exhaust conditions of the steam turbine match the process requirements.

The thermal energy absorbed from the high pressure steam by the backpressure steam turbine must be provided by burning an additional 46 MMBtu/hr of fuel in the ductburner. The efficient thermal energy conversion of the steam turbine combined with the efficient fuel utilization of the ductburner provide an additional 11,450 kw at an incremental cogeneration heat rate of 4017 Btu/kwh. This improves the cogeneration heat rate from 5042 Btu/kwh to 4937 Btu/kwh as shown in Table 9-2.

The steam exhaust flow of 664,000 pph at 600 psig/750°F together with the lower pressure HRSG steam production, completely satisfies the process requirements. Although the equipment and erection cost of the steam turbine is high in comparison to the previously discussed cogeneration options, it provides an increase in net present value to $142.53 million, an IRR of 52.12%, and a simple payback of 2.02 years. The incremental investment of $3.57 million results in a simple payback of 1.22 years. This configuration yields the highest net present value, thereby resulting in the largest increase in net worth to the plant owner.

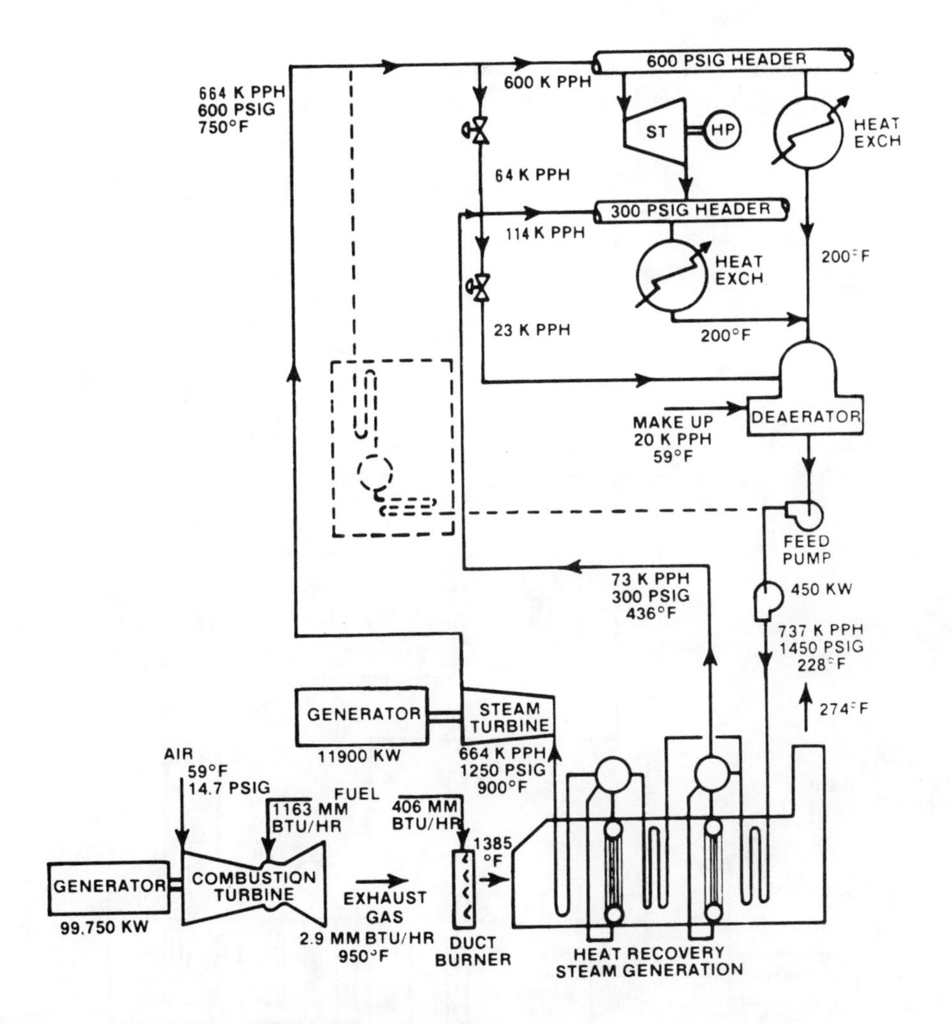


**Figure 9-6. Configuration D—Cogeneration Power Increase Backpressure Steam Turbine**

## Cogeneration Power Increase—Supercharging

Thus far in the discussion of cogeneration options, the combustion turbine component has remained unchanged; however, both increased power output and exhaust gas flow can be achieved by pressurizing ambient air prior to its entry in the combustion turbine compressor. This scheme, commonly referred to as supercharging, is shown in Configuration E (see Figure 9-7).

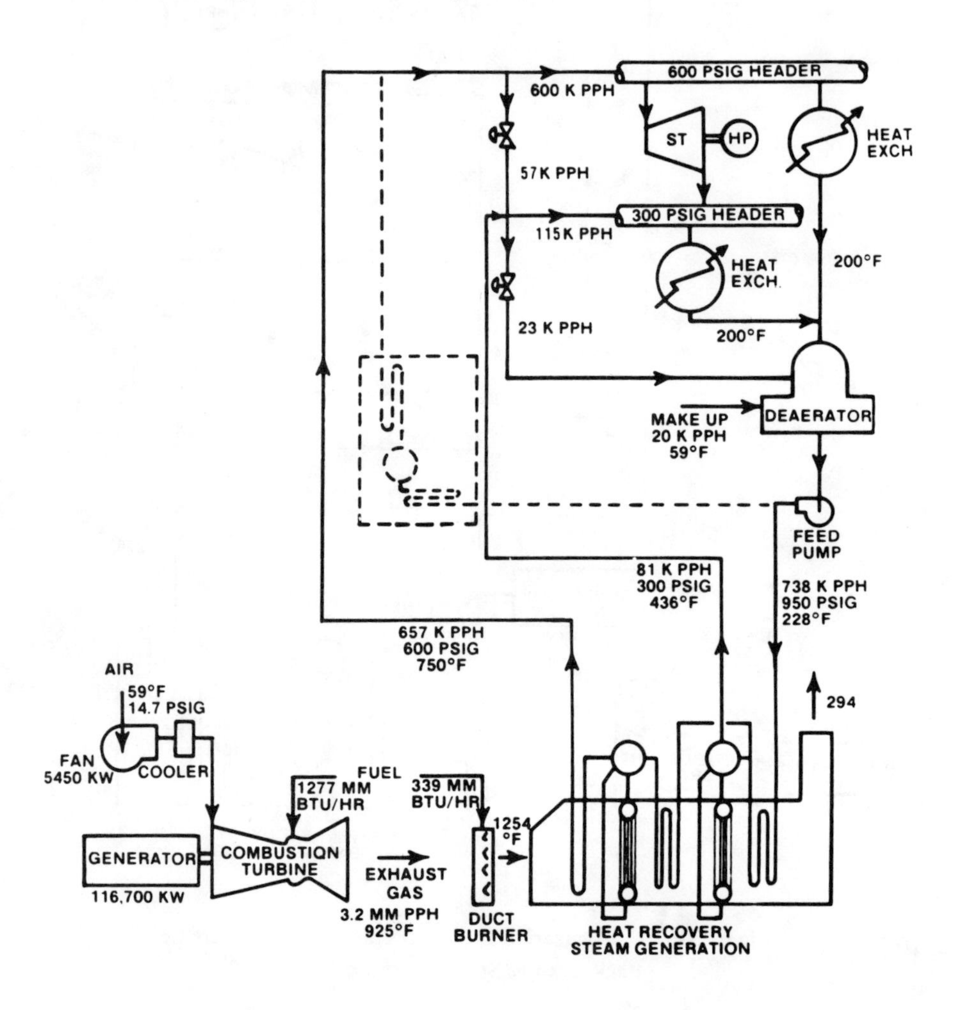


**Figure 9-7. Cogeneration Power Increase—Supercharging**

This configuration involves the addition of a large motor-driven centrifugal fan and an evaporative cooler to absorb the heat of compression. This additional apparatus results in an increase in combustion turbine power output of 11,500 kw (after deducting the fan power usage) and an increase in fuel usage of 114 million Btu/hr. The exhaust gas flow has been increased by 10% so that the process steam requirements can be met by supplementally firing a multiple pressure level HRSG to 1,254°F (compared to 1,337°F for Configuration C).

The higher exhaust flow reduces the HRSG heat recovery efficiency by 2% and increases the stack temperature 20°F so that the additional power is generated at a heat rate somewhat higher than that of the backpressure steam turbine (Configuration D); however, the equipment and erection cost associated with supercharging is only about half that of the steam turbine on a $/kw basis.

As seen in Table 9-3, the NPV associated with supercharging yields a net present value of $136.13 million, an IRR of 51.38% and a simple payback of 2.03 years. The $2.0 million investment in the supercharging system results in an incremental simple payback of 1.23 years. Note should be made that supercharging is most advantageously applied when steam demand is greater than that of the typical process used in our example so that increased exhaust gas flow is required to allow the HRSG to completely satisfy the process requirements.

## Other Considerations

The major objective of this chapter has been to discuss various options that affect the efficiency of combustion turbine cogeneration plants and the commensurate net worth impact to the firm (NPV). In particular, five cycles have been applied to a representative industrial process. Many other requirements and/or constraints could also be present. In this section, some of the more common process restraints are noted and briefly discussed.

### *Separating Electrical Power Generation From Steam Demand*

In some cases it may be desirable to insure constant electrical power production even though the process steam demand is reduced or eliminated. This situation may arise because the process electrical load is independent of its heat load or because of contractual requirements with the local utility. This situation can best be handled by incorporating a bypass exhaust stack between the combustion turbine and the HRSG. The transition and stack would be equipped with dampers so that all or a portion of the exhaust gas could be diverted as required.

The opposite situation may also prevail; that is, it may be necessary to insure constant steam production when the electrical power production is reduced or eliminated. For this case, the ductburners could be oversized so that the HRSG inlet gas temperature and the commensurate steam production can be maintained as the combustion turbine load and exhaust gas temperatures are lowered.

#### *Incorporating a Backup Source of Steam Generation*

Certain processes require an uninterruptible source of steam supply. For a combustion turbine cogeneration plant, this may be accomplished in several ways:

- Conventional boilers can be maintained in hot standby mode providing a backup to both the combustion turbine and HRSG.
- The HRSG can be equipped with fresh air firing capability backing up the combustion turbine.
- Two combustion turbines can be installed each with its own "oversized" HRSG equipped with ductburners. Normally the process steam demand would be satisfied by operating both CT/HRSGs without supplemental firing; however, either HRSG would be capable of meeting steam demand alone when fully fired.

### Other Factors

Other factors which may have to be considered include:

- Process steam demand transients.
- Utilization of process waste gas for cogeneration fuel.
- Environmental emission limitations and emission controls.

**Conclusions**

Combustion turbine-based cogeneration cycles can provide attractive investment options to energy intensive process industries, and can meet a wide range of process energy demands by judicious selection of cycle components.

Cogeneration cycle enhancements such as ductburners, additional HRSG pressure levels, backpressure steam turbines and supercharging can provide very attractive incremental investment options which can improve the profitability of the combustion turbine cogeneration project.

# Chapter 10

## Cogeneration: Planning for Nonconventional Technologies

***Beno Sternlicht***

Cogeneration/in-plant generation represents a very broad field. To illustrate the breadth of applications, this chapter presents the following four examples:

- Commercial/industrial heat pumps using gas/oil as an energy source
- Gas/liquid/vapor Waste Heat Recovery Systems to produce electrical/mechanical power and steam
- Heat Recovery Systems using solid waste to generate electrical power and steam
- Distributed Power using Indigenous Fuel Cogeneration

The state of technology and economic considerations are discussed, and the energy savings that can result from such energy cascading systems are presented.

### Appropriate Technology Perspective

Appropriate technology is simply technology that is appropriate for the problem. The level or degree of technology must be appropriate for the solution of the problem considering the location of the problem, the available resources, and the skills of the people. The size or scale of the technology must also be appropriate for the problem and it must be energy-efficient and economically and environmentally acceptable. In short, appropriate technology is often the antithesis of conventional wisdom of the American approach to problems.

## "Appropriate" Conversion Technologies

Consideration of appropriate energy conversion technologies should begin with the introduction of the Energy Cascading concept. Energy Cascading is simply the matching of the quality (temperature) of available energy with its use. This broader definition of energy conversion technologies includes concepts such as cogeneration, combined cycles, total energy, district heating, distributed power, etc. In Energy Cascading, the conversion process may simply be a series of thermal exchanges as well as thermal exchanges and power (usually, but not necessarily, electric) generation. Energy Cascading does not differentiate between bottoming or topping cycles. The only criterion is that the quality of energy match the need.

Application of the appropriate technology and Energy Cascading concepts to a number of potential energy conservation problem statements results in some interesting cogeneration concepts. These concepts involve the combining of technologies into cogeneration systems that are small, 3 kw–5 mw, and that are inherently energy-efficient. These concepts involve capital investment per kilowatt of capacity comparable to that for large central power stations, either coal or nuclear based.

The electric industry recognizes the potential energy conservation from the use of heat pumps and, therefore, is heavily promoting them. Heat pumps are not normally considered cogeneration systems because the power generation function is remote from and not integrated with the heat pumping function. Although it may be considered heresy, it is a fact that on-site thermally activated heat pumps are being developed under government and private funding. Market acceptance and technical feasibility of on-site thermally activated heat pumps are not that remote. Currently, modified total energy systems for on-site generation are being marketed and sold in New

York City and other parts of the country. Full integration of commercially available on-site generation systems with heat pumps is being tested in Europe and Japan. The next step of fully integrating an appropriate prime mover into an on-site thermally activated heat pump system is near.

Industrial thermal waste utilization is a clear form of energy cascading and cogeneration. The technical potential for power generation with Rankine bottoming cycles on the tail-end of industrial thermal discharges is approximately 3,000 to 4,000 mw. In terms relative to the electrical power generation industry, this capacity is small. Regarding the technical potential for classical cogeneration systems, it represents 5 to 10 percent of the projected conventional cogeneration potential in 1985. However, the scale, cost, returns, and constraints may be more advantageous for the bottoming cycles.

In some cases, considerations other than energy availability or cost can provide the economic driving potential for cogeneration. Industrial solid and liquid wastes must be disposed of in a safe and environmentally acceptable manner. The cost of this disposal is very large.

In the near future, some forms of acceptable disposal will not be available at any cost, e.g., landfills. An appropriate technology which simultaneously disposes of the industrial waste and generates useful thermal energy and power can be economically viable. The economics of these systems are driven not by energy savings, but by avoided disposal costs; however, the systems do conserve exhaustible energy sources.

For the past decade Mechanical Technology Inc. (MTI) has been developing Stirling engine technology. It is now applying this technology to cogeneration and other end uses. The Stirling engine uses external combustion. Thus, it has the advantage of being a multifuel engine using not only liquids and gases, but also solids and radiation.

In addition, the engine has high efficiency when compared to Otto, Brayton or Rankine cycles, and low emissions. Recently MITI in Japan also initiated the development of the Stirling engine for residential and commercial heat pumps and cogenerators. Their present development effort in this area centers on 3–30 kw systems. China has also made a commitment to the Stirling engine for cogeneration using coal and biomass as fuel.

## Heat Pumps

The energy conservation potential of heat pumps is well recognized. These devices seem to defy nature since they deliver more energy than they consume. This is true only if we consider the energy for which we must pay, the electric power to run the compressor and fans. However, the cogeneration concept can make the heat pump even more efficient in large commercial structures. Specifically, on-site thermally activated heat pumps can provide both power (shaft and electrical) and supplemental heat.

A cogeneration concept to provide electric power and building heat is known under the term district heating. District heating implies large-scale systems, possibly entire cities. Various problems with classical district heating systems have prompted the consideration of smaller scale projects centered upon a "community."

This concept of dispersed power generation integrated into a community's energy generating and consumption needs is under development by the Department of Energy under the term of Integrated Community Energy Systems (ICES). The concept of dispersing the power generation function has been developed to more readily integrate the thermal waste with thermal needs.

In other words, it is felt that a reduction in scale is more conducive to Energy Cascading. The selection of the proper scale of implementation will affect the selection of the appropriate technology and the ultimate energy conservation potential. The selection of the proper scale is dictated by the appropriate definition of a "community."

The appropriate definition of an energy community would be one that fully recognizes the legal, social and technical environment in which we live. Currently, the most common definition of an energy community for heating and cooling purposes is a single building. This is a specific case of the community energy system. Thus, in order to maximize energy conservation potential and minimize social dislocations, an energy cascaded cogeneration system should be considered for a single building.

This chapter will be restricted to a discussion of large commercial buildings, where there are often simultaneous heating and cooling loads and where some near term technologies are feasible. Further, since the heat pump is the most effective means of heating when there is an available source, the single building cogeneration system should be based upon a heat pump.

The power for the heat pump would be provided on-site by a conversion of fuel to mechanical power and/or electricity. The mechanical power or electricity would drive the heat pump compressors. The source for the heat pumps would be the hottest available source necessary to meet the load. This would include refrigeration condensers, ventilation air and outside air.

The on-site thermally activated heat pump would seek to maximize energy conservation by limiting operation to that which is necessary to meet the building heating and cooling loads and the resultant parasitic power demands. In this respect, on-site power generation for heat pumping is very different from Total Energy. Total Energy seeks a complete fuel substitution often in a very energy inefficient but economic manner due to relative fuel and electric costs.

The schematics shown in Figures 10-1 through 10-3 illustrate the conventional HVAC approach, energy conserving approach and on-site power generation or cogeneration approach for the winter heating balance. The cogeneration approach provides the total building heating and power for a portion of the HVAC compressor and fan loads with less fuel than the conventional or energy conservative approaches.

As with many concepts, this has some clear advantages as well as possible disadvantages. While this is accomplished by maintaining the current fuel mix, it would not result in the consumption of any more "scarce" fuels for heating than before. However, it would inhibit the switch to more abundant fuels through electrification.

The schematic in Figure 10-4 illustrates the heat pump cogeneration approach for the summer heat balance. During the summer or cooling season, it is not clear if this concept offers any societal advantages. This system might not save fuel in the summer since energy cascading cannot be effective when there is minimal need for internal heating (only hot water).

A comparison of fuel type and efficiency between the power generated on-site and a central peaking unit may show no fuel type or

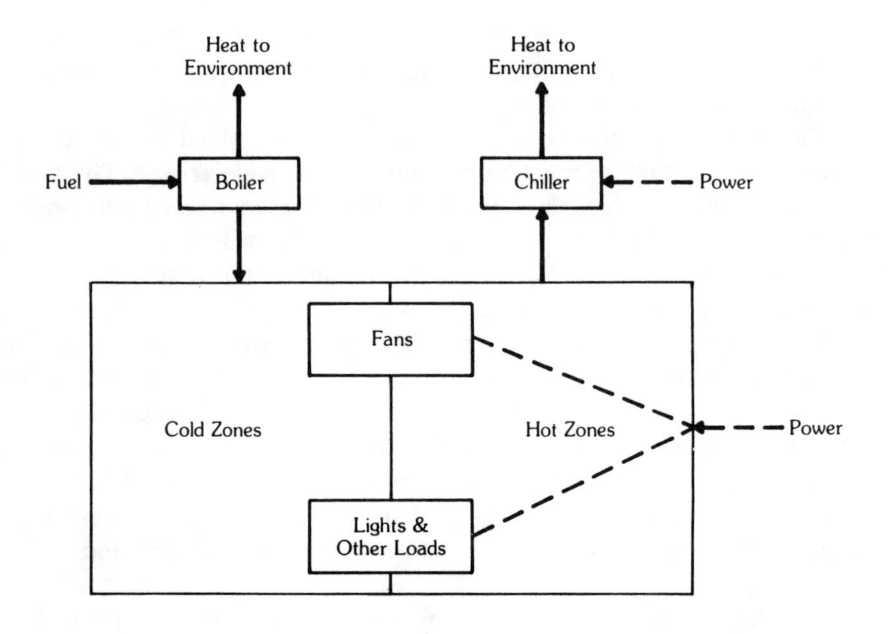


**Figure 10-1. Winter Heat Balance on a Large Building—Typical Approach**

efficiency advantage for either approach. Peaking plants tend to be less efficient than base plants and tend to use "scarce" fuels. However, an argument for an on-site thermally activated heat pump based upon marginal central station efficiencies is rather tenuous. The trade-off in capital investment may also lead to an indifference between approaches.

The argument for on-site thermally activated heat pumps must be based upon its overall fuel efficiency in producing shaft power for the refrigeration compressor relative to the central station approach. The relationship of the heating and cooling requirements of most commercial buildings is such that the cooling requirements dominate both the energy conservation potential as well as the economics.

As shown in Figure 10-5, the fuel to shaft power efficiency of the on-site system must have a fuel conversion efficiency of at least 25 percent. This efficiency must include all the steps directly in the conversion process, such as for a Rankine system the boiler, turbine, and

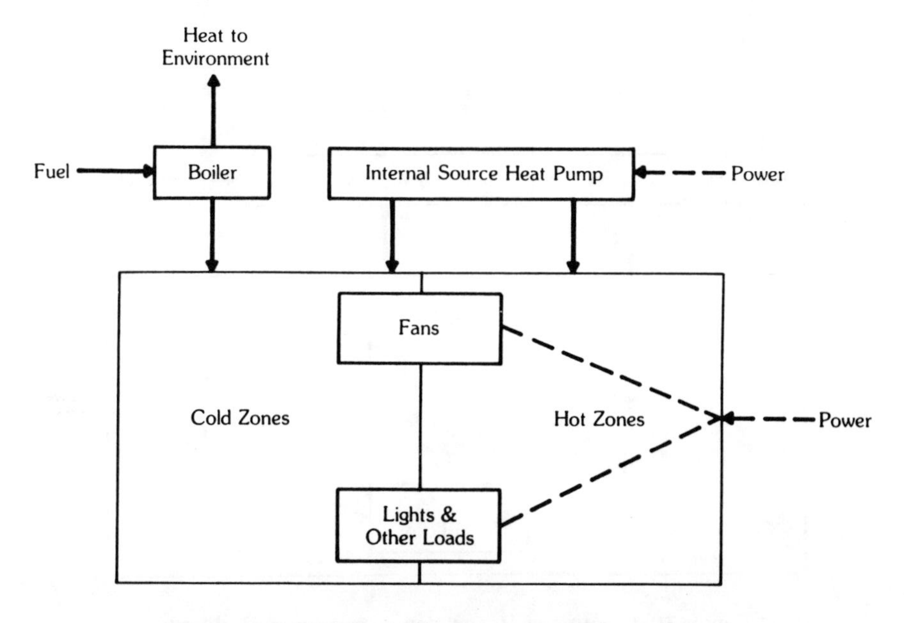


**Figure 10-2. Winter Heat Balance on a Large Building—Energy Conservative Approach**

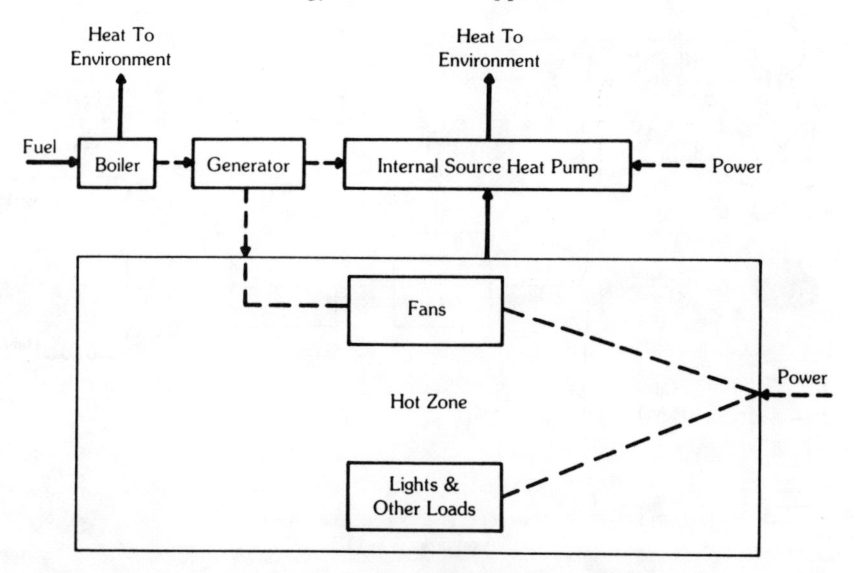


**Figure 10-3. Winter Heat Balance on a Large Building—Cogeneration Approach**

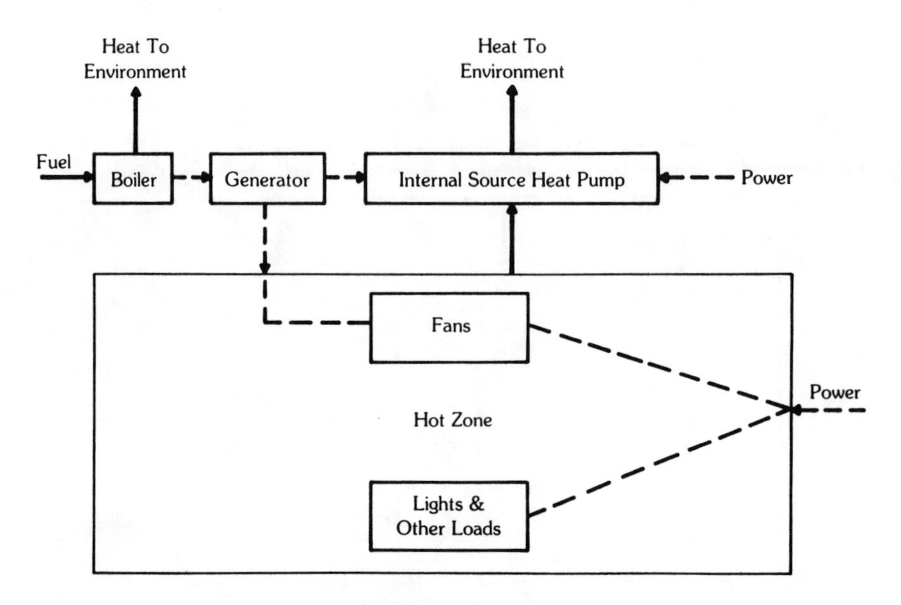


**Figure 10-4. Summer Heat Balance on a Large Building— Cogeneration Approach**

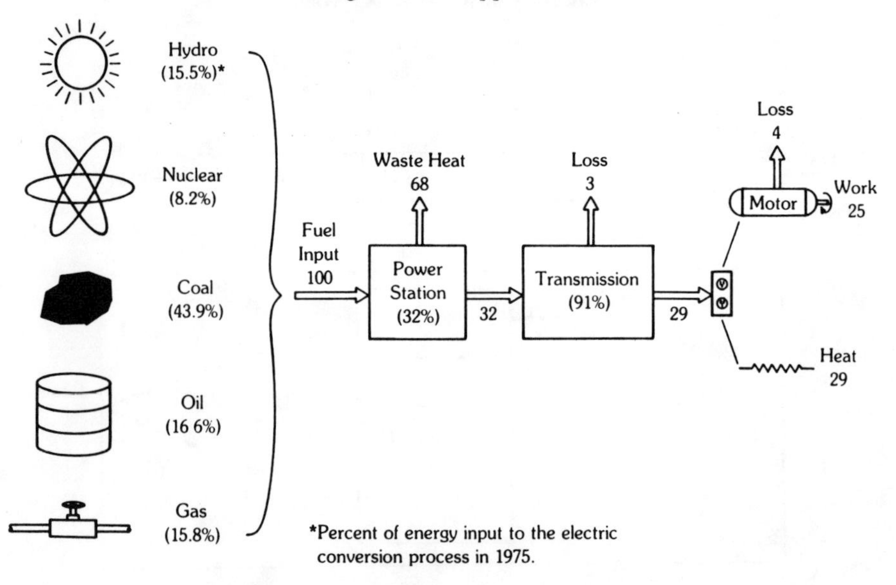


**Figure 10-5. Electric Conversion Process**

gear train, as well as parasitics, such as lube oil pumps, cooling tower, deaeration systems, etc. A number of small prime movers offer the potential of achieving at least the level of conversion efficiency. The prime mover cycle with the highest efficiency potential is the Stirling followed by Diesel, Brayton and Rankine cycles.

Regardless of which prime mover cycle is ultimately found appropriate, the concept offers the potential of energy conservation. Information gathered to date on a cogeneration heat pump system shows that an individual building owner will usually use less fuel energy but always pay less to operate this cogeneration heat pump system (Table 10-1).

**Table 10-1. Operating Costs with an On-Site Thermally Activated Heat Pump Compared to Current Standard Practice**

| | | *Annual Operating Costs* | |
|---|---|---|---|
| | *Base* | | *Heat Pump* |
| Gas | $ 3,993 | | $15,030 |
| Electric | | | |
| Base | 18,922 | | 3,529 |
| Demand | 7,609 | | 5,345 |
| Total | $30,524 | | $23,904 |
| Savings | | $6,620 | |

## Bottoming Cycles

Cogeneration with topping cycles in very large scales (10 to 100 mw) is the subject of much attention. While the potential is viewed as being very large, it has not and may not be realized. Another form of cogeneration with bottoming cycles with industrial waste heat has not received as much attention. The potential technical market is smaller than that for topping cycles.

However, the economics of the bottoming cycles relative to the topping cycles can be more competitive. Furthermore, the energy conservation potential of the bottoming cycle is much greater than with topping cycles since *no* incremental fuel must be consumed to generate the power.

On March 1, 1979, MTI officially commissioned the first installation of a Rankine bottoming cycle on a municipal electric utility diesel plant. The overall installation is shown in Figure 10-6. The Rankine system can operate off either or both of two Diesel engines rated at 5.5 mw. The full power output of the Rankine system is 500 kw with 60 percent flow from both engines.

When the Rankine cycle is operated from one, the rated output drops to about 400 kw. The Rankine system adds about 8 percent of the rated power of the base prime mover without any additional fuel consumption. This means that the overall heat rate of the station is decreased from about 8,600 Btu/kw (40 percent efficiency) to about 8,000 Btu/kw (43 percent efficiency).

The Rankine system installed at Rockville Centre is unique in that it is a Binary Rankine cycle. The Binary Rankine cycle uses two working fluids as illustrated in Figure 10-7. The use of two working fluids has a number of advantages as well as some disadvantages.

The advantages include the use of well-known and acceptable working fluids that represent no threat; they are nontoxic and nonflammable. The system operates above atmospheric pressure and thus avoids some design and operating problems. The use of a low temperature organic fluid permits the full utilization of low temperature condensing sources. Disadvantages include a more complicated machinery and control system. The use of seals presents an area of potential reliability problem.

The largest potential energy conservation market is expected to be in the process industries where large amounts of heat are rejected at 300°F and below (Figure 10-8). In these applications, an organic cycle based on a fluorocarbon is the most attractive system.

The economics of Rankine cycles in industrial applications are beginning to be attractive in many process industry locations. There is an economic regime for Rankine cycles which depends on the temperature, type (condensing, gas or liquid), and size (heat flow) of the stream. Temperature affects the overall thermodynamic efficiency

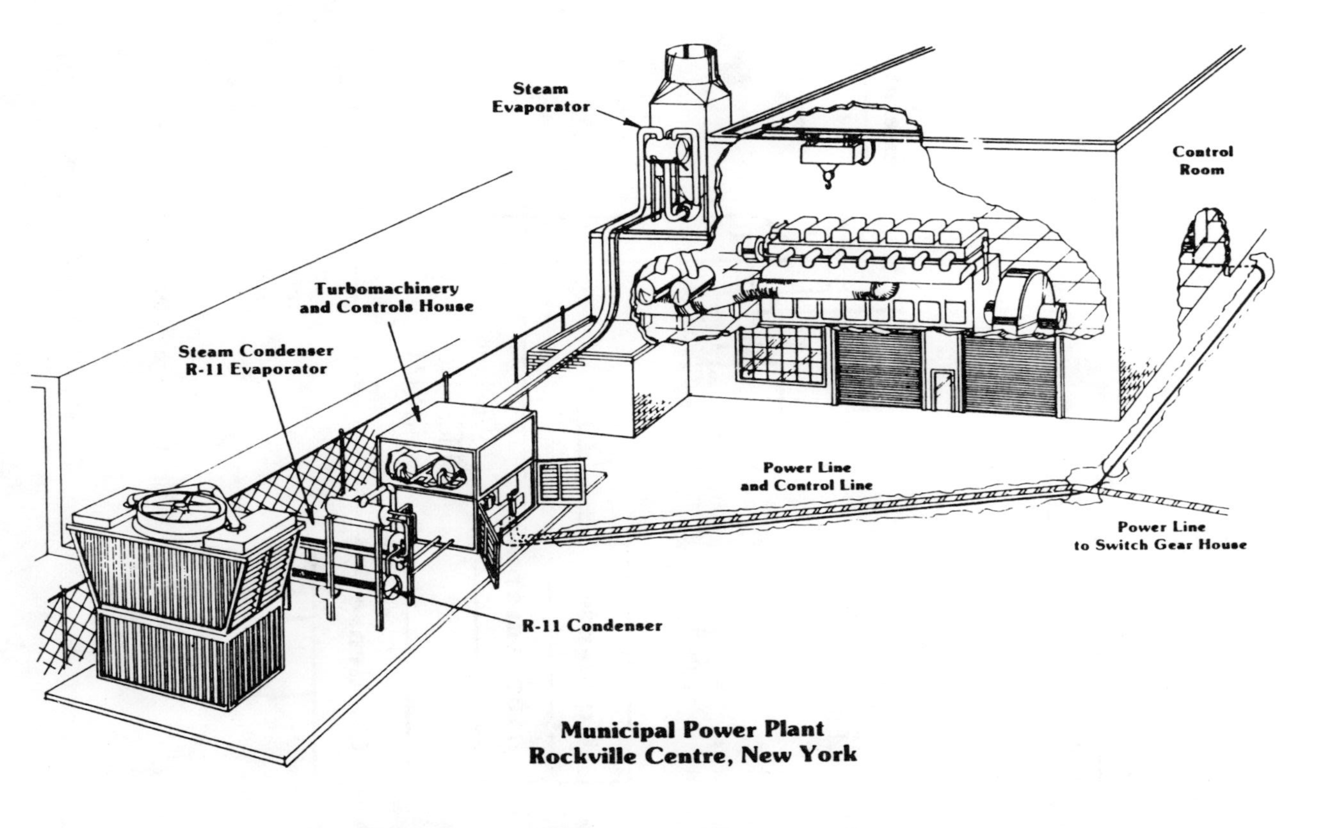


Figure 10-6. General Arrangement of MTI Binary Rankine Cycle System for Waste Heat Recovery/Electric Power Generator

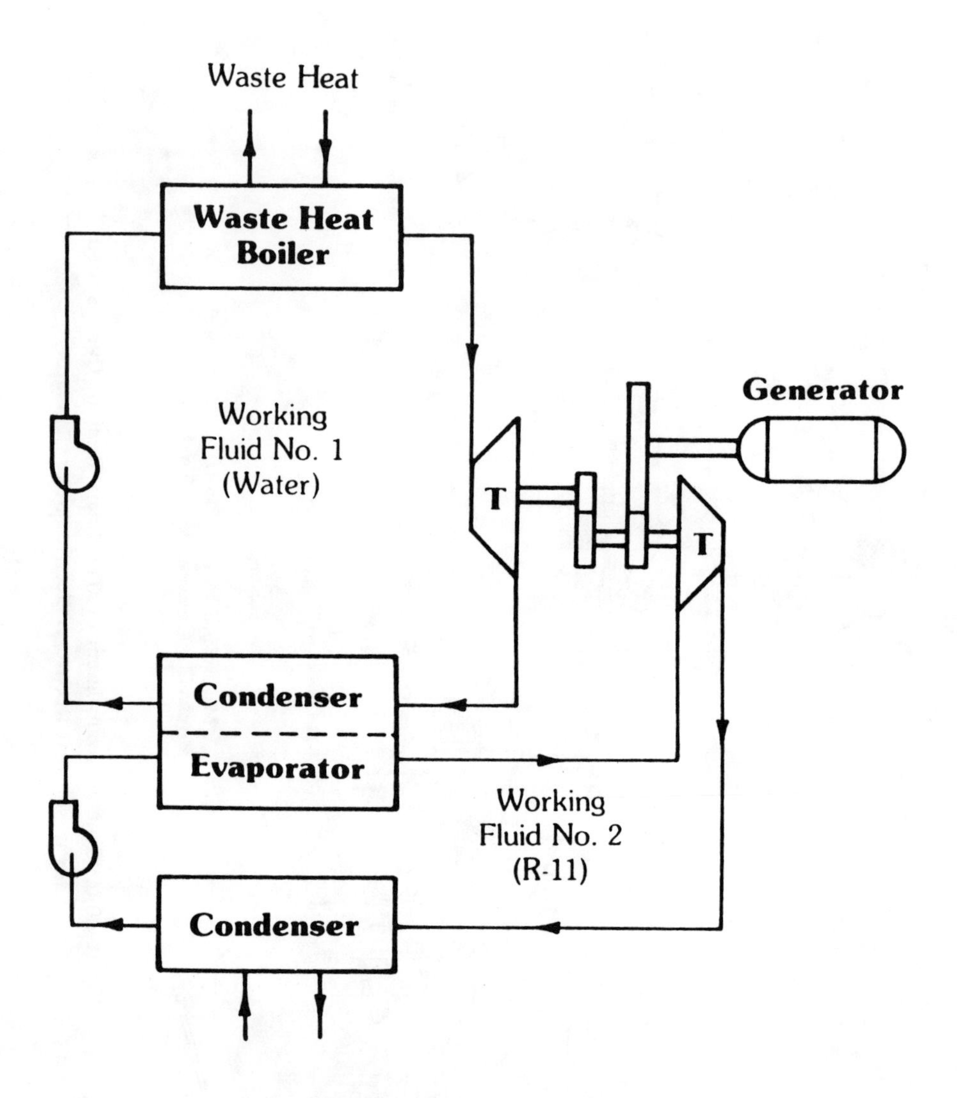


Figure 10-7. Binary Cycle Simplified

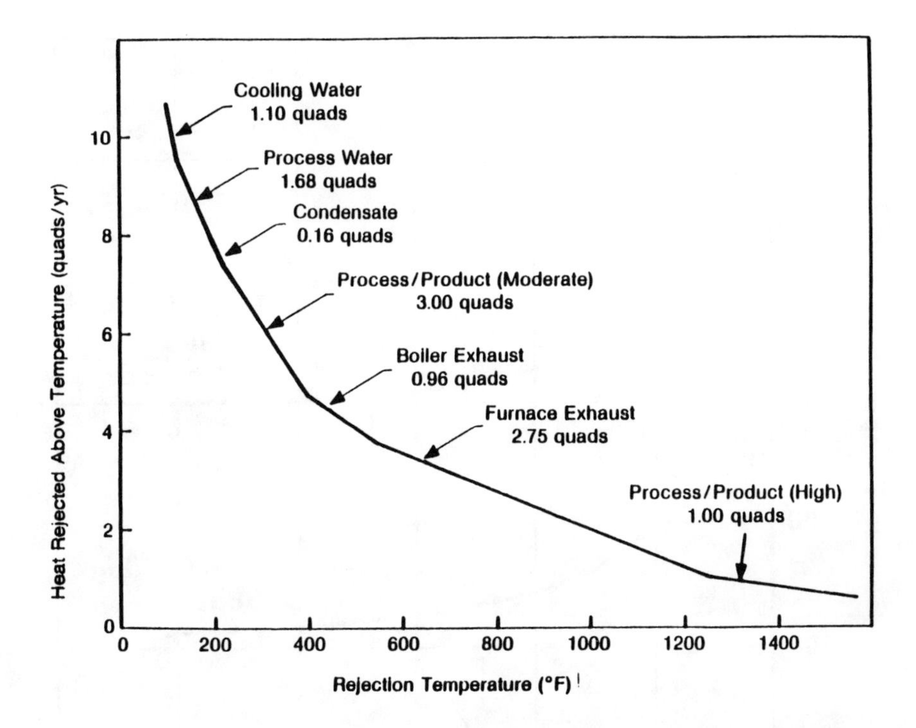


**Figure 10-8. Distribution of Industrial Waste Heat Flows**

*Source: "Industrial Applications Study," DOE Report HCP/T 2862-01, March 1978*

of the cycle and governs the amount of power that can be recovered for a given stream size and amount of machinery.

The type of stream impacts the size, type and cost of heat exchangers. Stream size affects the total amount of power that can be recovered at a given temperature. More power at a given temperature spreads the fixed cost over a larger base, and thus reduces the machinery cost per kw of capacity.

This economic trade-off is shown in Figure 10-9 for both a condensing vapor waste heat source and a hot gas source. The bottom line for any application is the economic attractiveness in an actual industrial environment. Two options for an organic Rankine bottom-

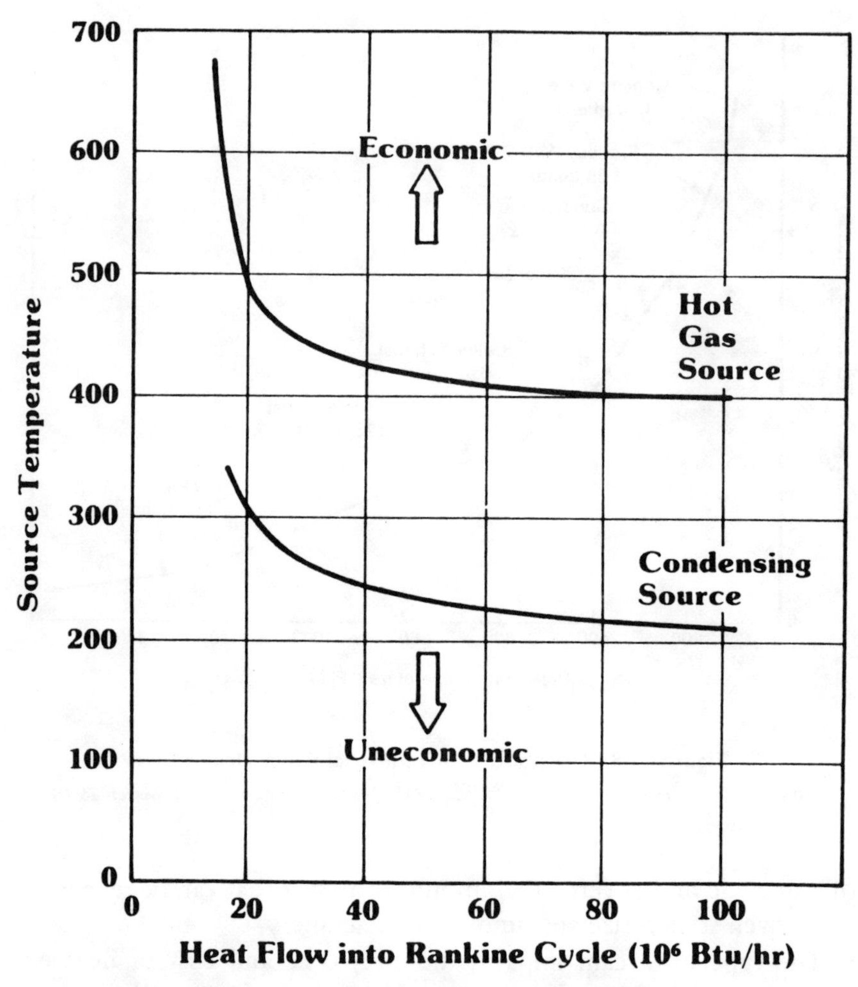


Figure 10-9. Economic Applications for Bottoming Cycles

ing cycle in a refinery are examined in Table 10-2. The installed costs (approximately $1600/kw) of these options are typical of what would be expected in a good application. The economic analysis of these options shows that even with costs of $1600/kw, the investment should be attractive to industrial firms, with payback period about 3-5 years and discounted cash flow (DCF) rate of return about 26–27%.

**Table 10-2. Application of Organic Rankine Cycles in a Refinery**
*Overhead Condensing System for a Fractionator*

| | *Option A* | *Option B* |
|---|---|---|
| *Heat Source Conditions* | | |
| • Type—Combination of: | (Linear condensing hydrocarbon and liquid gasoline streams) | |
| • Inlet Temp, °F | 299 | 315 |
| • Exit Temp, °F | 200 | 200 |
| • Heat available million Btu/hr | 43.6 | 56.0 |
| *Cooling Water Inlet Temp, °F* | 80 | 80 |
| *Power Recovery Cycle* | | |
| • Heat extracted million Btu/hr | 40.9 | 51.8 |
| • Source discharge temp, °F | 206 | 209 |
| • Turbine inlet temp, °F | 230 | 240 |
| • Net cycle efficiency % | 13 | 13 |
| • kw output, net | 1354 | 1781 |
| *Heat Exchanger Data* | | |
| • Source heat exchanger surface, $ft^2$ | 14,178 | 19,452 |
| • Condenser heat exchanger surface, $ft^2$ | 29,612 | 46,228 |
| • Cooling water flow, gpm | 4828 | 6075 |
| *Cost (Thousand $)* | | |
| • T-G Module | $ 666 | $ 811 |
| • Source heat exchangers | 142 | 194 |
| • Condenser | 317 | 370 |
| • Installation | 1125 | 1375 |
| – Total | $2250 | $2750 |
| – Unit cost, $/kw | $1662 | $1544 |

While the largest potential for bottoming cycles is at low temperatures with organic Rankine cycles, there is waste heat available at high temperatures which can also be recovered by bottoming cycles. At waste heat source temperatures above 600°F, the preferred approach is a steam Rankine cycle. At very high source temperatures (e.g., 1600°F and above), the use of Brayton cycles has been investigated.

A steam Rankine bottoming system fueled by the exhaust of 12,500 hp gas turbine was put on line last year at the Pacific Gas & Electric Company's Gerber, California pipeline compressor station. This unit (5,300 hp) recovers 4 mw of electrical power. This is equivalent to over 42% of the gas turbine nameplate rating power with a net reduction in heat rate from 10,500 Btu/hphr simple cycle to 7,500 Btu combined cycle.

The installed cost for this system including engineering and equipment is roughly $2500/kw. This includes the waste heat recovery boiler, steam turbine generator, air-cooled condenser, building and auxiliaries. At this cost, the system shows good discounted cash flow. It is estimated that replication of this system should cost in the order of $1900/kw.

Bottoming cycles are technically and economically feasible. Only actual in-plant demonstrations of these facts now inhibit their acceptance. This hurdle needs to be removed by actual plant demonstration programs.

## Industrial Waste Disposal

Waste disposal requirements presently pose as serious a threat to many companies as does the energy situation. These companies are faced with the problem that current disposal techniques may not be available in the very near future. The alternatives are to move to more expensive and, oftentimes, more energy intensive waste disposal techniques or close the doors to their plants.

The economic penalties imposed by the former alternative may ultimately lead to the latter. An appropriate cogeneration technology would combine the disposal of the waste with useful energy production. The fuel value of these wastes is rather large as indicated in the selected listing shown in Table 10-3. The heat value of these wastes is comparable to or better than some low-grade fossil fuels such as lignite. However, like low-grade fossil fuels, these wastes are hard to burn and present pollution problems.

**Table 10-3. Fuel Potential of Industrial Wastes for Cogeneration Systems**

| *Industry* | | *Fuel Value of Waste (Quads)* |
|---|---|---|
| Wood Products | | 0.26 |
| Commercial Refuse | | 0.26 |
| Pulp and Paper | | 0.18 |
| Sugar | | 0.11 |
| Industrial Chemicals | | 0.05 |
| | Total | 0.86 |

There is considerable effort to develop technology to combust low-grade fuels such as municipal wastes, high sulfur coals, and low-ranked coals. Most of the effort is directed toward large facilities, well over 20 mw. However, a survey of industrial plant waste streams indicates that most plants have waste disposal requirement sufficient to support the generation of less than 2 mw. This disparity in scale plus the prior lack of need has resulted in industrial waste being hauled away and buried.

Few small incineration companies are making small incinerators for hospitals and other small commercial applications. None of these cogenerate, although some now provide steam.

If the technology were economically available, a better use and disposal of these industrial wastes would be on-site incineration with simultaneous steam and power generation. This type of small scale, combined waste disposal and cogeneration facility has been investigated in considerable detail by MTI. We call this system concept CO-DYNE.™

Recent advances in two technologies have now brought the concept of industrial-sized units closer to favorable economics. First, small fluid-bed combustion technology is being pursued by a number of companies. These fluidized beds offer the potential of cleanly and efficiently incinerating wastes in units of 1–2 mw. Small packaged Rankine systems being developed by MTI for bottoming cycle applications indicate that these power systems can be economic. The CO-DYNE system joins these technologies in a pre-engineered, pre-packaged modular plant which burns industrial wastes in a fluidized bed to generate steam and power.

A sketch of the CO-DYNE concept is shown in Figure 10-10. This sketch shows the system taking in 100 tons per day of industrial wastes and producing 2 mw of electrical power. A split between pure electric generation and steam-electric generation depends on the plant's need for steam and electricity. For example, either 2 mw of power or 28,000 lb/hr of process steam and 50 kw of electricity may be produced.

The economics of the CO-DYNE system are governed by the fuel disposal credit. Figure 10-11 shows the sensitivity of the economic viability of the CO-DYNE concept to fuel costs. Within the current range of electric power costs (30 to 60 mils/kwh), only industrial wastes that have fuel disposal credit can be economically justified. However, the technology is appropriate since it safely disposes of industrial wastes and economically produces a useful output.

## Distributed Power

In the United States, our home energy needs are generally served by two utility systems. The electric utility supplies electrical power for our lights, appliances, etc., and the gas utility supplies gas for our hot air furnaces, water heaters, etc. In some cases, electricity and gas are sold by a single utility company. With the availability of cost-effective cogenerators, all of our house energy needs could be supplied with one energy conversion system.

This would result in a very significant energy saving. On a life cycle cost basis, this could present a very good business opportunity. The distributed cogenerators could be owned by the electric or gas

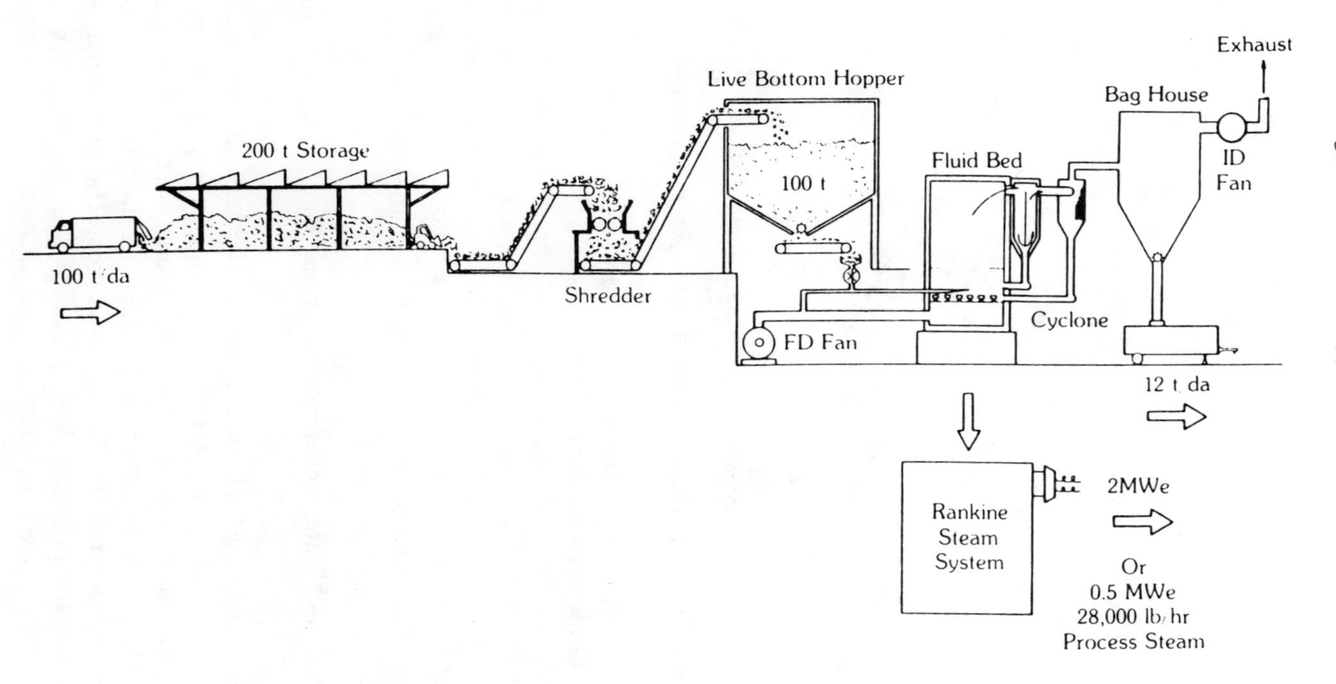


**Figure 10-10. Overall CO-DYNE™ System Concept**

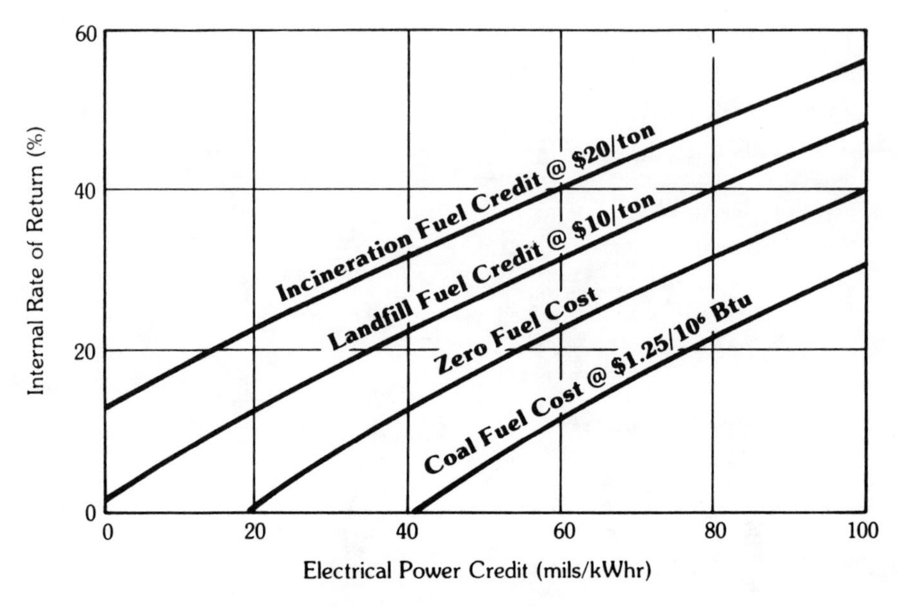


**Figure 10-11. The Economics of CO-DYNE™ Power System**

utility or a third party financing company. These entities in turn would sell Btu and kw to the homeowner.

The homeowner could also own such energy conversion equipment, but it will, of course, require him to make the initial investment. What are the roadblocks to this obviously sound cogeneration system and business opportunity? There are several:

- Availability of cost-effective "small" generation systems
- Utility management attitude
- Federal and state government attitude, e.g., taxes, incentives, regulation.

With the rising cost of fuel, greater acceptance of cogeneration systems, and advances in several technology areas, this enormous business opportunity has attracted very strong interest by a number of companies in this country and abroad. On a short-term basis, Otto and Diesel cogenerators are being considered. The disadvantages of these systems are primarily low reliability and relatively low efficiency for electric power generation.

Japan, Italy and Germany are already demonstrating some of these systems. In this country, Diesel cogeneration systems are being introduced in schools, hospitals and commercial establishments. The fuel cell promoted by United Technologies Corporation has been considered for residential applications, but it appears to be too expensive. It seems to be better suited for larger size systems.

The best candidate residential energy conversion unit for small power levels in the range of 3–50 kw appears to be the Stirling engine. This engine is highly efficient, and low in emissions and noise. The free piston Stirling engine (FPSE) offers also high reliability and long life since it is a process fluid lubricated hermetic engine. In addition, this is an external combustion engine that can use any fuel including solids. Recently, Japan and China have made a commitment to this technology for distributed power.

**Conclusions**

It is obvious that certain components and portions of Energy Cascading are technically within the state-of-the-art. Cogeneration in large plants and district heating with back pressure turbines has been practiced for decades. However, to achieve maximum efficiency of energy utilization, as much energy cascading as technically and economically possible must be carried out.

This requires that new technologies be developed and demonstrated in order to give industry technically viable alternatives from which they can choose. It also requires that incentives to use state-of-the-art equipment, with only small improvement in technology, be avoided. This comment may sound contradictory, but we must remember that we have another scarce resource—*CAPITAL.*

It is important to point out that the investment required to put a new product into the marketplace is often very large. Industry must recover its investment before it is prepared to make another significant product change that requires similar investment. A consumer, likewise, is not prepared to obsolete the equipment unless he/she has obtained a reasonable value from its use.

We, therefore, must optimize the time for introduction of a new product based on realistic projection of technology advances, investment required and national objectives. The impact of the timing of the introduction of energy conservation approaches and their relative conservation potential is illustrated in Figure 10-12. Even though the conservation impact of a current technology can start sooner, its cumulative impact may be less than the potential of an advanced approach.

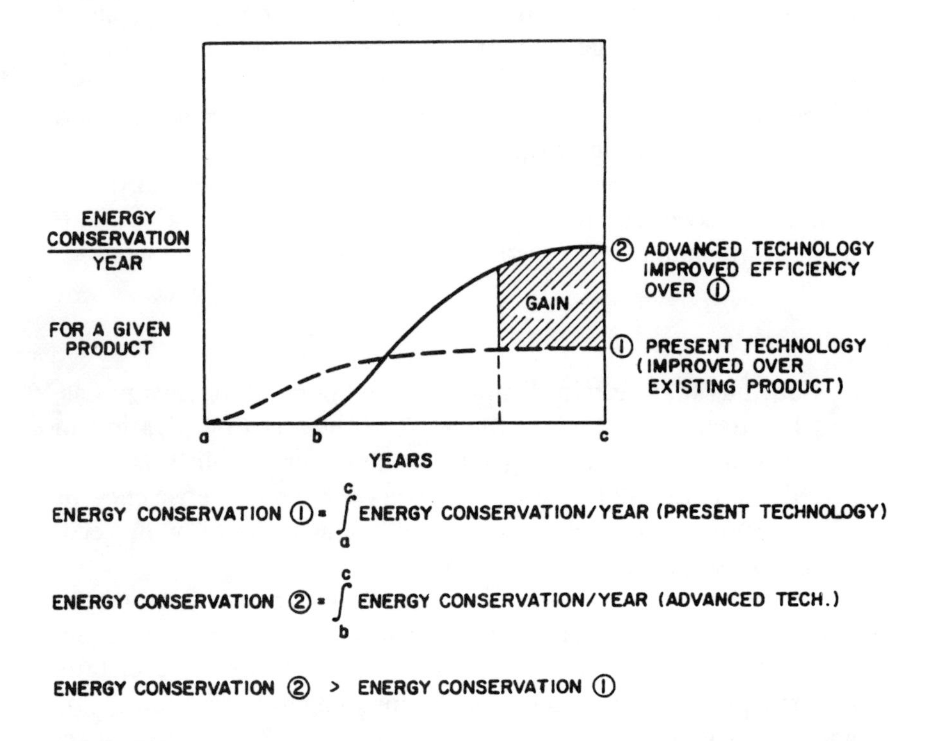


**Figure 10-12. Relative Energy Impacts with Present and Advanced Technology**

Thus, from a long-range national perspective, it may be better to discourage the adoption of current technology in favor of a new technology at a later date. This simple observation is often not appreciated; yet it presents the key rationale for the need of a vigorous and continuing technology effort. Hopefully, it also cautions against incentives and regulations that are poorly assessed.

# Chapter 11

# Cogeneration Fuels in the 1980s

**Robert G. Uhler**

Although the Public Utility Regulatory Act (PURPA), the Federal Energy Regulatory Commission (FERC) and, recently, the Supreme Court have all given cogeneration a legal boost, it is ultimately the cold objectivity of the marketplace that will determine the long-run fate of cogeneration. Certainly, technology will play a major role, but that too can be reduced to economics. Focusing more narrowly still, fuel costs are of particular importance in determining the future of cogeneration.

National Economic Research Associates, Inc. (NERA) has assessed the availability and has estimated the price of three prime cogeneration fuels in the 1980s: No. 6 residual fuel oil, No. 2 distillate oil, and natural gas. Coal, of course, is also an important fuel for cogeneration but is not discussed here. In general, cogenerators need to compare their fuel costs with their local electric utility's avoided energy costs.

This chapter analyzes part of that comparison—the prices of three fluid fuels for cogeneration. Each fuel has economic linkages with the others as competitors not just for cogeneration uses but in other energy markets as well. For example, the prices of No. 6 oil and No. 2 oil are importantly determined by the availability and price of crude oil which also affects the price of natural gas. In brief, crude oil is the dominant energy commodity throughout the world. And, as such, it is the linchpin for fuel prices. Therefore, a discussion of the market outlook for crude oil comes first, followed by a look at each of the three cogeneration fuels.

### Crude Oil

U.S. crude oil prices, of course, are a reflection of world crude prices, which in turn remain influenced by OPEC policies and actions. The rise in Free World oil consumption ended in 1979 at a peak level of 50 million barrels a day (see Figure 11-1). Since then consumption has fallen to a level of 40 million barrels a day in 1982 and even lower in the first quarter of 1983. This drop reflected conservation and substitution (as the normal economic response to higher oil prices), and economic recession (in part induced by those prices). At the same time, crude supply has also responded in normal economic fashion to higher oil prices: although overall Free World production has declined, non-OPEC output has risen (see Figures 11-2 and 11-3).

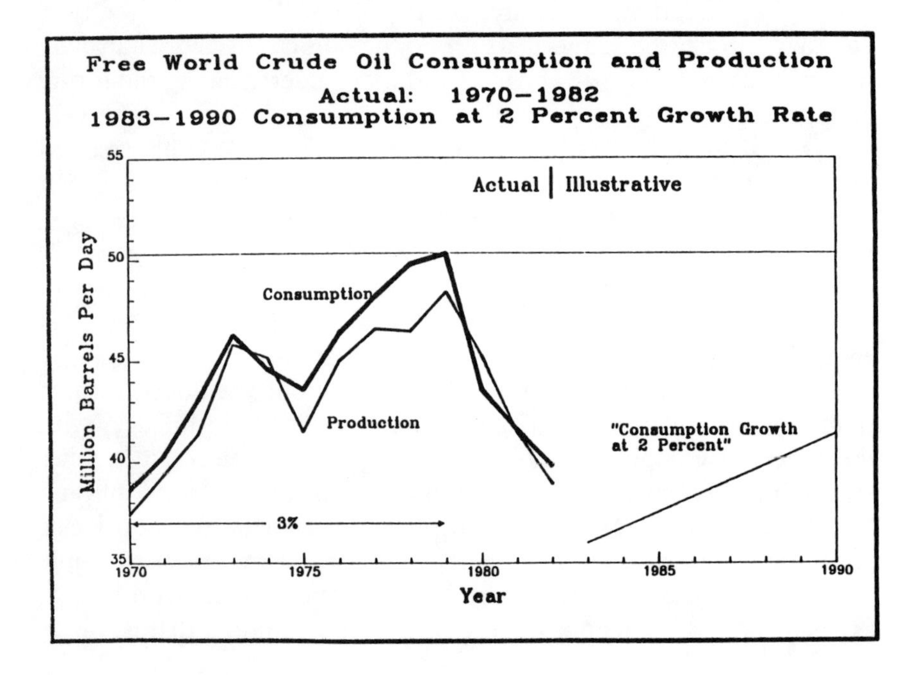


Figure 11-1

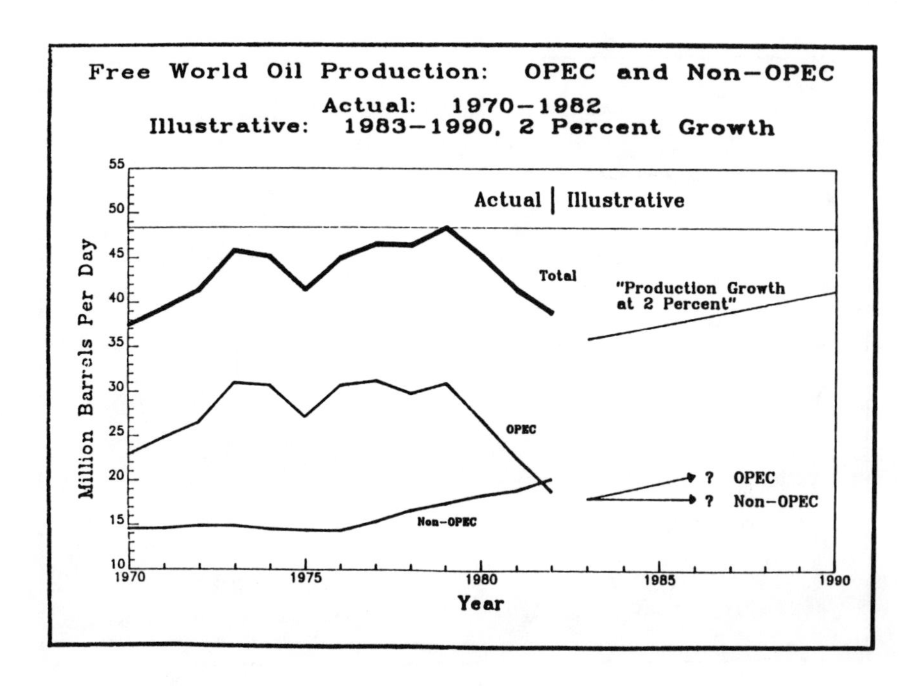


Figure 11-2

This combination of declining demand and increasing non-OPEC production put growing pressure on OPEC to adjust its pricing policy in recognition of these changing circumstances. Yet until March 1983, OPEC was unable to agree on any price-output action despite repeated attempts to do so. Facing the grim prospect of total market collapse, OPEC finally fixed a price of $29 per barrel for Arabian Light crude (the "marker" price on which the prices of OPEC crudes of higher or lower quality are based). OPEC also agreed to establish a mechanism for monitoring the production quotas necessary to maintain that price.

Like all OPEC decisions of the past two years, the March agreement was a gamble, counting on world economic recovery and an end to the drawdown of stocks to revive oil demand. So far, the gamble appears to be paying off. The awful consequences for OPEC members of a market collapse have motivated a previously unattain-

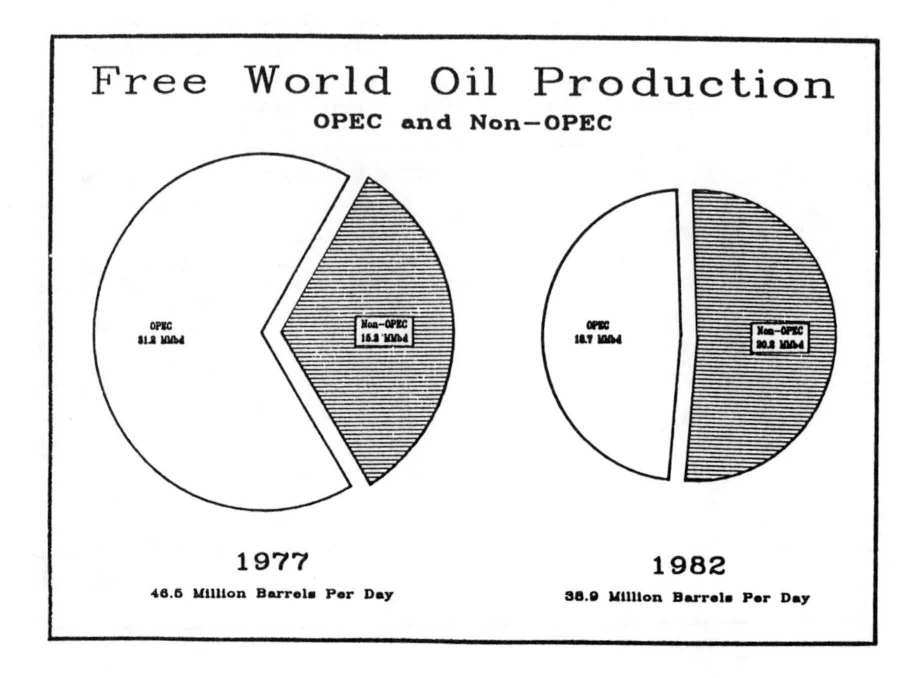


**Figure 11-3**

able and uncharacteristic discipline in adherence to quotas. The large non-OPEC exporters, the United Kingdom and Mexico, have unofficially cooperated in setting their own prices in line with the OPEC price structure. Equally important, a definite economic recovery has begun in the United States. As things now stand, the odds favor the avoidance of a price collapse in world oil markets.

Economic recovery during 1983 steadied oil prices but is not likely to permit any increase in real crude prices from the present level. In the ensuing years, continued economic recovery should raise the demand for crude sufficiently to allow the OPEC countries to produce at their quotas without any trouble but still not put upward pressure on real prices (oil prices in current dollars could increase at the rate of inflation). For most of those countries, present quotas are below those nations' physical capacity to produce; thus the desire to supply more than the quotas will put a brake on any oil price increase. Indeed, one OPEC minister has stated that the $29 price structure will be maintained through 1985.

Statements like this from OPEC officials should not be taken at face value, but the pronouncement does recognize marketplace realities. Much of the depressing effect on demand for crude oil caused by the oil price increases of the 1970s is permanent. For example, automobile efficiency in the United States will never again fall back to what it was; indeed, the turnover of the auto fleet will alone cause the average efficiency to continue to rise although the renewed consumer interest in "big cars" may moderate this process.

Moreover, insulation and other energy-saving retrofits in existing houses and structures will certainly not be torn out, and all new construction will continue to have higher heating/cooling efficiency levels, in many instances mandated by law. Still further, substitution away from oil in new facilities by electric utilities and industry in general will not be reversed (unless there is complete failure in attempts to solve the natural gas price muddle—see below), and new industrial plants built as economic recovery progresses will be even more energy efficient.

So whatever the pace of future economic growth, the rate of increase in oil demand will be less than that. Since 1980 half of the decline in oil consumption has been attributed to the recession and half to conservation. Thus, if future economic growth is at, say, 4 percent annually (which would be fairly healthy), the increase in Free World oil demand would be more like 2 percent per year (see Figure 11-4). It would take a growth rate in oil demand much higher than 2 percent (in fact, over 5 percent) to regain the 1979 consumption level by 1990; yet to get such an increase in oil consumption, the economic growth rate would have to be something like 10 percent per year.

Obviously, any strengthening of the demand for crude oil will tend to firm up its price, but that price effect will be tempered by circumstances on the supply side. To begin with, Saudi Arabia, having borne the brunt of the production cutbacks during the recent lean years, will wish to get back to its previous output level of 10 million barrels a day as quickly as possible from its current production of about 5 million barrels. But other countries will also wish to

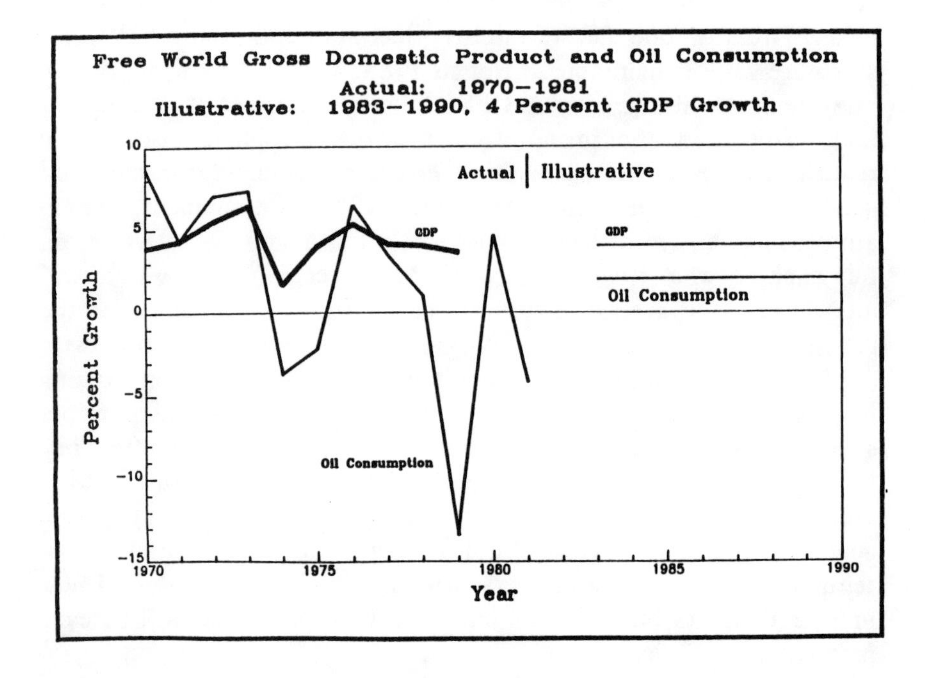


Figure 11-4

take advantage of market strength; Nigeria, Venezuela, Indonesia, Libya and Mexico would all like to export more oil.

As economic recovery is completed and "normal" economic growth is resumed in the world, the most that is likely to happen is a restoration of the loss in real price since 1981. Consider this: current OPEC production is down more than 13 million barrels a day from its 1977 peak of 31 million barrels a day. Thus, world demand will have to increase by some 40 percent before there will be any pressure on OPEC capacity. Saudi Arabia could, of course, forego increasing its production and speed up the rise in price. More likely, the Saudis will follow a middle course. They will not increase production as fast as crude oil demand grows, thus holding the price at the present level. They will increase production slowly, behind the rate of demand increase, thus permitting the price of oil to increase.

If economic recovery is complete within a few years, NERA

believes this would mean that the previous Saudi marker price of $34 per barrel is likely to be restored in the latter 1980s, perhaps as early as 1986 (see Figure 11-5). Note, however, that $34 is a current dollar price. Inflation, it is safe to assume, will have continued in the meantime, so that when the $34 level is reattained, the price of marker crude will actually be lower than that in real dollars. Thus, in 1986, the price of oil in 1982 dollars will still be about $28 per barrel. Thereafter, the price of oil (in current dollars) will probably do no more than keep pace with inflation for the remainder of the decade. Or, to put it another way: after the $34 price (in current dollars) is restored, the price of oil in real dollars will remain unchanged, at least through 1990. Figure 11-6 converts the dollar per barrel values to an MMBtu (million Btu) basis. Three reasons for our view on crude prices are summarized below.

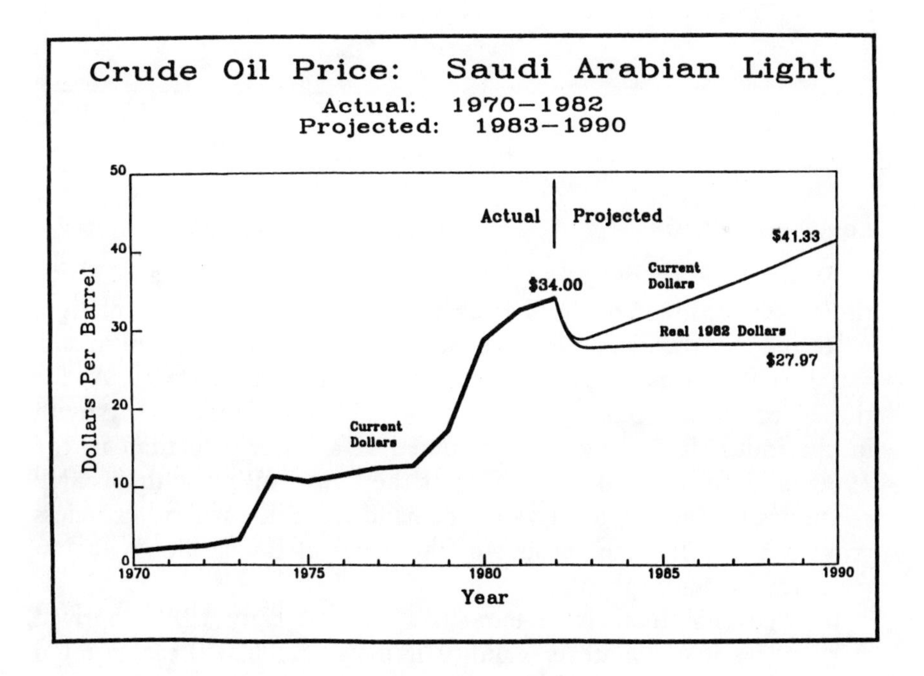


Figure 11-5

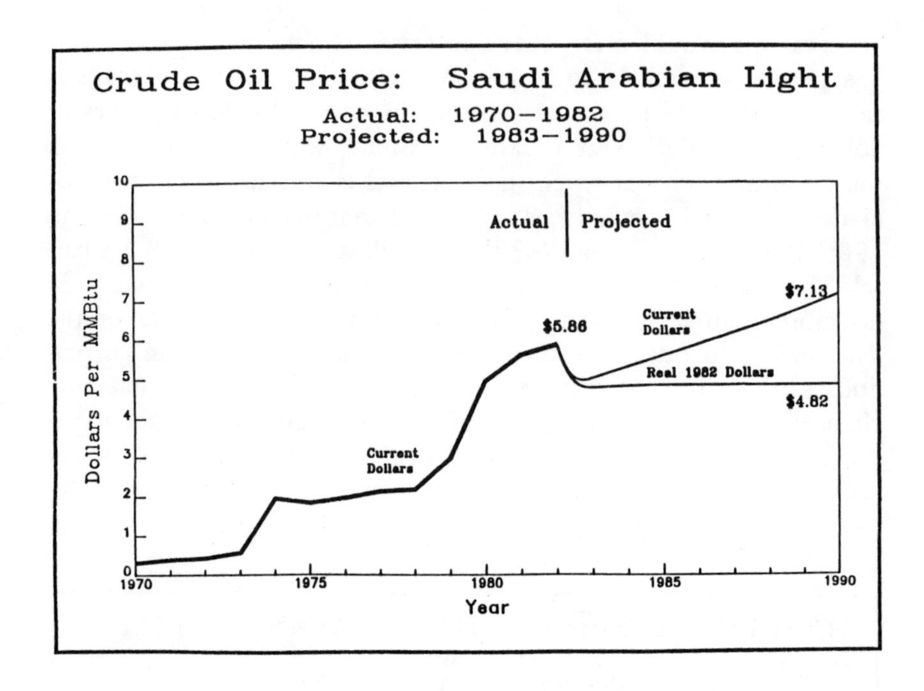


Figure 11-6

*Long-Term Demand*

At the time of the $5.00 price cut in 1983, oil was about five times as expensive in real dollars as it was prior to the OPEC embargo. After the cut, the incentive to conserve and substitute was still strong, and it will continue to exist in the industrialized countries. As for the Third World, some analysts argue that even if economic growth in the industrialized world is resumed at a lower rate than in the 1960s and early 1970s, and even if the use of oil per unit of GNP continues to decline, total world demand for crude will nevertheless grow at a healthy rate because of the needs of the developing countries. This seems doubtful.

International financial events in 1982 and early 1983 deprived this argument of whatever validity it may have had. Bear in mind that for those developing countries (indeed, for all countries other than the U.S.), the real price of oil was *not* falling prior to the March

1983 price break. OPEC oil is sold in U.S. dollars, and the appreciation of the dollar relative to almost all other currencies has maintained or even increased the real price of oil in terms of those currencies.

Thus, the nonproducers of oil in the Third World have been hit with the double blow of high oil prices on imports of that commodity and reduced demand for their exports due to the world recession. While their export earnings have fallen drastically from the combination of smaller volumes and lower prices, their hard currency needs to pay for oil imports have either remained constant or have risen. Foreign borrowings that were probably excessive in any event are now beyond the ability of many Third World countries to service.

Calamity has been avoided by recourse to the International Monetary Fund (IMF). The IMF, however, makes its loans under the condition that the borrowing country put its financial house in order (i.e., requiring that the local government impose domestic measures which tend to lower incomes and demand). Thus, given the financial extremities of many Third World countries, reducing their external debt load under IMF guidance will at the very least dampen those nations' domestic growth and more likely result in a contraction of those countries' economies. In a word, the putative growth potential in the demand for oil by the Third World will not be realized for many years.

### *Long-Term Supply*

On the long-term oil supply side consider, first of all, Iran and Iraq (see Figure 11-7). Their war will probably be resolved in one way or another by 1990. Whoever is the winner, or if the military stalemate is recognized in a peace settlement, both countries will be eager to resume exporting crude oil up to their respective capacities. Under the late Shah, Iran's output reached 5.7 million barrels a day, but its reported oil exports are now only about 2.1 million. Iraq previously produced 3.5 million barrels a day but is now down to 1.0 million.

Resolution of the war, therefore, could release some 6 million barrels a day of additional supply on the market; and both countries, if they wished, could expand capacity beyond the levels that existed prior to the Iranian revolution and the war. These 6 million barrels–

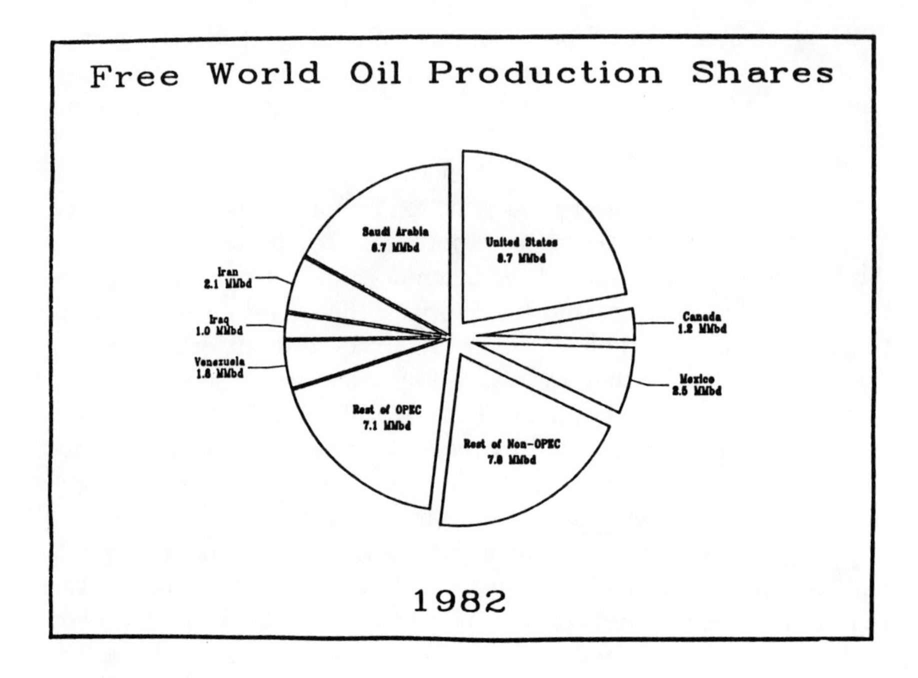


**Figure 11-7**

15 percent of today's Free World market—would be in addition to approximately 12 million barrels a day of other excess OPEC capacity.

There is also the potential for increase in non-OPEC supply. The most important source, because it is not only the largest but also the most certain (in the physical sense), is Mexico. On the basis of known reserves, Mexico's productive capacity could be increased from its present 3 million barrels a day to 10 million (although not, to be sure, by 1990). A level of 5 million barrels a day by 1990 is feasible, and the economic forces motivating such an increase are undeniable. Oil exports provide Mexico the best means of handling its enormous foreign debt problem.

Other areas in which productive capacity could conceivably increase between now and 1990 are the North Sea and the Alaskan North Slope. Discoveries continue to be made in the North Sea, and the British government has recently encouraged the development of previously marginal fields by sharply reducing the tax rate on North Sea operators. In Alaska, several large discoveries have already been made since the Prudhoe Bay find. Part of the resultant new capacity will go to offset the decline in Prudhoe Bay production, which will begin before 1990. Moreover, it would be physically possible, if the North Slope operators so desired, to bring enough additional productive capacity into being that it would be necessary to raise the capacity of the Trans-Alaska Pipeline System (TAPS) from its present level of 1.6 million barrels a day to the 2.0 million barrels for which it is designed.

### *OPEC Policy*

As pointed out above, OPEC policy will support no increase in the real dollar price of oil at least for the remainder of the decade. The OPEC price increase of 1979 was a serious blunder that worked against the cartel's own long-term interest. That short-sighted action greatly intensified and accelerated the response of both demand and supply to the previous cumulative price increases, resulting in today's oil market weakness. The worldwide recession of the early 1980s has only magnified OPEC's problems, which would have existed in any event.

In recent months, the more reasonable policymakers in OPEC have publicly acknowledged this pricing mistake. Sheikh Yamani has stated that the long-term goal of Saudi Arabia is to maintain the real price of oil. There is no longer any talk of raising oil prices faster than the rate of inflation. In this instance, the Saudis have the full power to force what they want as OPEC policy on the other members.

Prior to the March 1983 meeting, the Saudis were not able to impose their desire to reduce the price because of the fear of triggering a price war. In contrast, to keep the price of oil from rising, all the Saudis have to do is increase production. At one time during the 1970s, their target for productive capacity was 12.5 million barrels a day; so if the Saudis want to keep the price of oil from increasing faster than inflation, they have the complete power to bring it about.

Unless OPEC were willing to do its catch-up very slowly, it would run the risk of restimulating the same economic forces of demand

and supply that led to its recent difficulties. The Saudis think this risk and its consequences are great enough to justify forgetting about the recent real price loss. The Saudis have the very long term in view; they do not wish to jeopardize permanently the market for their oil in the 21st century.

NERA concludes that there are no strong forces that would tend to increase the price of crude oil in real terms during the 1980s. Over the longer term there are many combinations of slight modifications in market forces and political factors that would make truly flat real prices in the 1990s unlikely. Taking this into account, the probability of flat prices or a rate of increase of about one percent or so per year in the real price of oil are equally likely. All of the foregoing, of course, is predicated on the assumption of a surprise-free future. Another Arab-Israeli war, an Iranian-style revolution in Saudi Arabia, or similar possible but unpredictable political events could have a wide range of effects on oil supply, world demand and price.

In brief, NERA believes that the reappearance of repeated large jumps in real crude oil prices, such as characterized the 1970s, is remote (Figure 11-8 shows refiner acquisition costs of crude oil on an MMBtu basis). There should be no upward pressure on the real prices of petroleum products during the remainder of this decade, and at most a modest tendency for real prices to increase during the 1990s. This stability in the price of the dominant energy commodity will mean, in turn, a constraint on the tendency for the real price of other fuels to increase. Finally, the economics of cogeneration are both enhanced and worsened. On the one hand, with no increase in the real prices of fuels, the cost of producing energy—both electric and thermal—using cogeneration will not increase; on the other, the greater efficiency of cogeneration will be a lesser advantage if fuels are less expensive.

### No. 6 Oil (Residual Fuel Oil)

As noted above, crude oil prices are the ultimate determinant of petroleum product prices. For the lighter oil products this is a reasonable generalization—the premium in their prices over the crude price has fluctuated within a fairly narrow range—but for residual oil

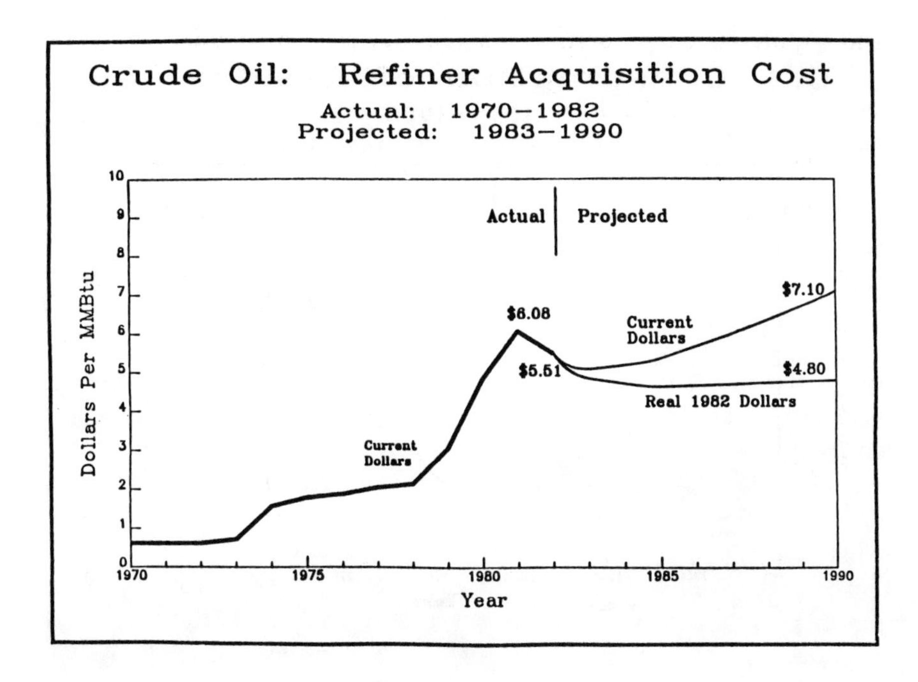


Figure 11-8

(resid), the discount in recent years has varied by as much as 50 percent of the crude price. Matters are further complicated by changes in the demand for resid, in part because of sulfur content considerations and, on the supply side, changes in refinery practices.

*The Demand for Resid*

Consider first the demand for resid. Between the mid-1960s and 1977 (see Figure 11-9), the trend of annual resid consumption was upward; since then there has been a trend reversal—demand in 1983 was approximately 46 percent below the 1977 level. Much of the decline in 1981 and 1982, however, obviously reflects recession conditions, which are temporary. Economic recovery should, therefore, bring with it a resumed growth in resid demand. But the recent low levels also reflect the cumulative response of consumption (i.e., demand elasticity) to the increase in real price since 1973.

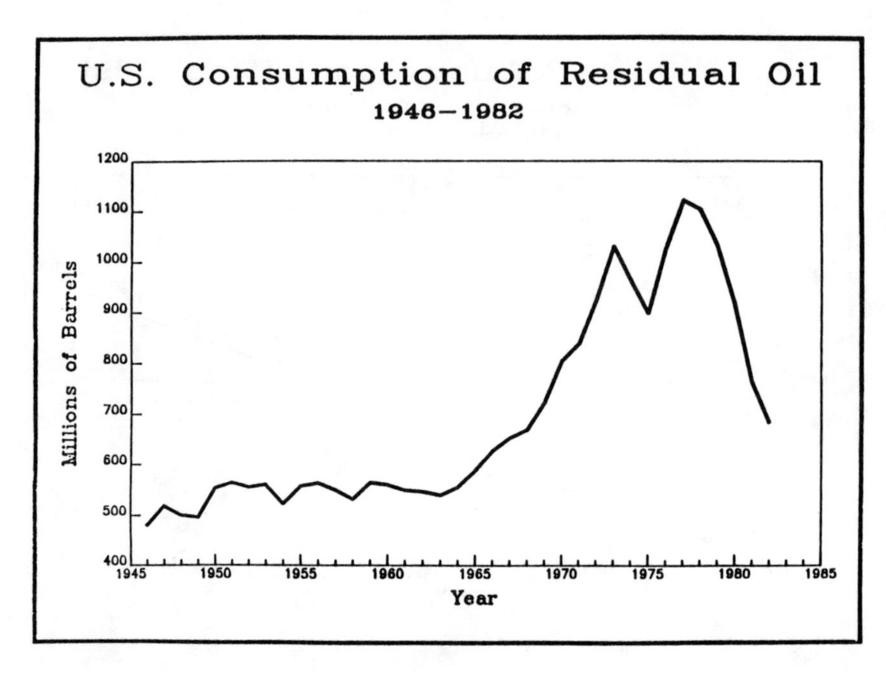


Figure 11-9

This has resulted in lower use of resid per dollar of GNP, which suggests that economic recovery will not bring a proportionate rise in resid use. The increase that actually occurs will depend on the relative contributions of recession and elasticity to the recent decline in consumption. Somewhat less than half of the decline in all energy use in the U.S. has been attributed to recession.*

*The Supply of Resid*

As for the supply side, Figure 11-10 shows the yield of resid as a percentage of the product barrel at U.S. refineries for the period since World War II. The persistent downtrend through 1970 reflects the fact that traditionally, the refiner has sold resid for less than the ac-

*E. Hirst, *et al.*, "Recent Changes in U.S. Energy Consumption: What Happened and Why," Oak Ridge National Laboratory, February 1983.

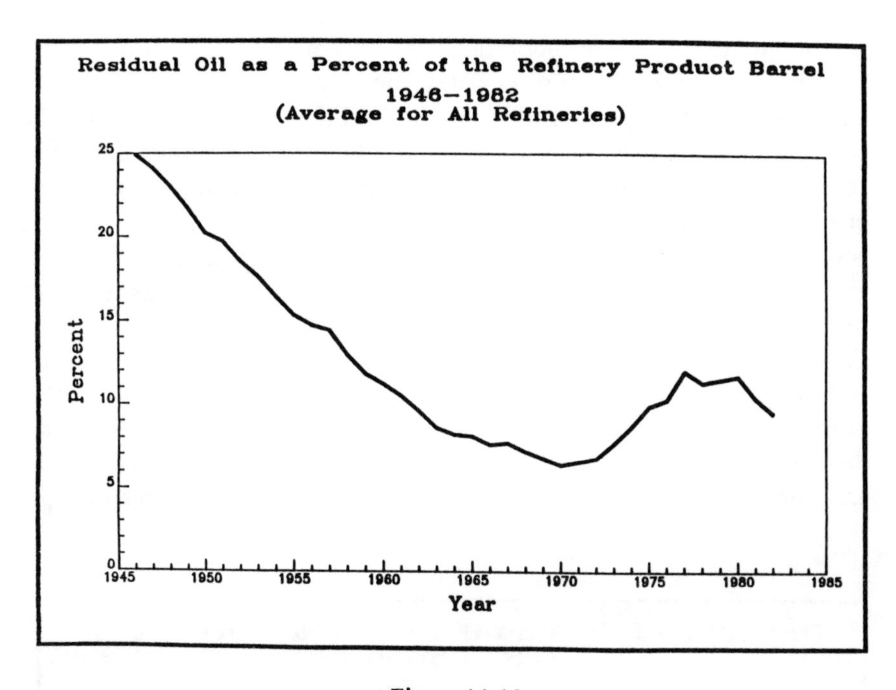


Figure 11-10

quisition cost of crude. During that same period, the growth in the demand for gasoline and distillates and the move to higher octane ratings in gasoline spurred refiners to increase the yield of the profitable lighter end of the crude barrel.

The trend reversal in resid yield between 1970 and 1971 is due, at least in part, to the stimulus provided by Federal law through the entitlements program. This encouraged the operation and construction of small refineries designed for straight-run distillation and consequent large resid yields. The decontrol of oil prices in January 1981 and the resultant phasing out of the entitlements program led to the closing of small refineries, coming as it did along with a general slackening of demand because of the recession. At the same time, larger refiners began to upgrade their facilities to handle larger proportions of heavier crude in their feedstock, as such crudes became increasingly available at appropriate discounts in the world market.

The upgrading of refinery facilities to convert resid into lighter products has accelerated since 1980; between now and 1986 "resid destruction capacity" in the U.S. will double. Moreover, overall U.S. refinery capacity has been declining in the face of reduced demand. In 1982, total capacity dropped by 9 percent, much of it older refineries that produced resid. Thus the domestic availability of resid is resuming its previous long-term downtrend. Historically, as shown in Figure 11-11 consumers have been able to meet their needs through imports, and a comparison of Figures 11-10 and 11-11 shows that there is a high inverse correlation between imports and domestic resid yields.

But there is another complication. The move away from resid production in this country is paralleled in refineries around the world, and for the same reasons. Heavy crude is an increasing proportion of

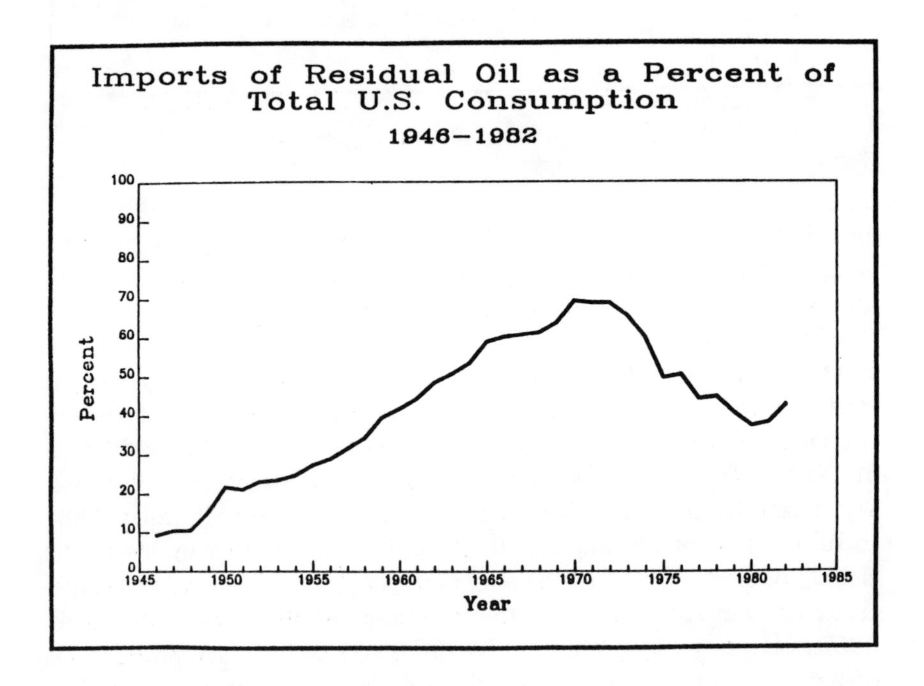


Figure 11-11

total supply, and as the industrialized countries try to reduce dependence on oil imports in reaction to the experience of the 1970s, resid consumption elsewhere in the world is also declining. European refiners expect a reduction in the proportion of resid in the product mix from 30 percent to only 21 percent in the next 10 years. By 1985, two-thirds of the refineries will have completed upgrading programs.

This increase in conversion capacity will stimulate what has been a minor element in the resid market—its purchase by refiners as feedstock. This, however, should not have any significant effect on the availability of resid by boiler fuel users. For one thing, the alternative will continue to be the purchase of heavy crude, with its attractive gravity discount. In early 1983, for example, prior to the OPEC realignment of oil prices in March, Saudi Arabian Light crude (34° gravity) had a posted price of $34 per barrel, while Saudi Heavy (27°) sold at $31. The relative prices of the various crudes also take into account sulfur content and geographical location.

Thus, the price of resid will tend to be kept under the price of crude by responses in both supply and demand. On the supply side, a resid price higher than crude would cause Venezuela, for example, to sell resid rather than crude. On the demand side, some boiler fuel users would burn crude directly rather than pay a higher price for resid. The current price of resid is low relative to the present crude price because world demand for resid has been declining faster than the decline in resid availability. Continued installation of conversion capacity, however, will further reduce that availability. Thus, the discount in the price of resid relative to crude may narrow somewhat over the next five years.

In summary, resid prices (see Figure 11-12)—on average—peaked in 1981 at $5.17 per MMBtu and are now near their short-run low—say about $4.50 per MMBtu in 1983 in current dollars (differences in sulfur content and regional disparities also need to be considered). NERA projects the real price of resid (delivered to large users) to be relatively flat for the rest of the 1980s—about $4.00 per MMBtu in 1982 dollars.

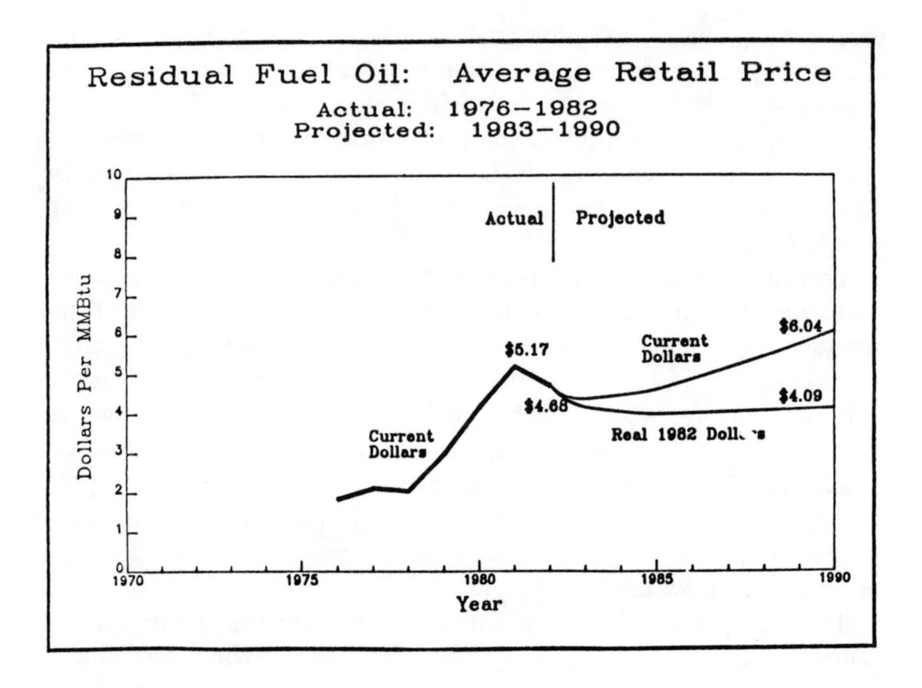


Figure 11-12

## No. 2 Oil (Distillate Fuel Oil)

Because diesel fuels—used in diesel engine cogeneration installations—are similar to "heating oils," a separate discussion of diesel fuel price trends is not necessary. Instead, we focus on the outlook for No. 2 (distillate or home heating) oil. This analysis is straightforward: an ample supply of crude and adequate domestic refining capacity will result in an ample supply of No. 2 oil. Moreover, middle distillates (i.e., refined products such as diesel fuel and No. 2 oil) will increase as a percentage of the total product barrel. Although there may be some difficulty in achieving the necessary yield from an average lower quality of crude feedstock to refineries,* on balance, the

***General Accounting Office, Potential Middle Distillate Supply and Demand 1982–1990,* December 17, 1982, p. iv.

rate of decline in feedstock quality should be gradual enough to permit the industry to cope with the change in quality.

Conceivably, the growth in demand for diesel fuel for vehicles could affect the availability and price of No. 2 oil. Until the softness in gasoline markets of the past 18 months, sales of diesel-powered automobiles and diesel fuel were growing rapidly. The effect of lower gasoline prices, however, was swift and direct. In the first half of the 1982–1983 model year, sales of diesel cars were 48 percent below the previous year.*

Since the end of price controls on oil in January 1981, the retail price of No. 2 heating oil paid by homeowners and small commercial or industrial users has fluctuated between 40 and 55 percent higher than the refiner acquisition cost of crude. These movements reflect such things as the lag between the receipt of crude, processing it and selling it; the level of No. 2 stocks at refineries; and market circumstances at the time of sale, which can affect the refiner's and distributor's margins.

There appears to be nothing that would change this general price relationship for the remainder of this decade. Accordingly, using the illustrative date of 1986 as the time of reattainment of the $34 price for marker crude oil, the retail price of No. 2 heating oil is expected to be about $1.19 per gallon at that time (in current dollars). In real terms (i.e., 1982 dollars), the retail price of heating oil will hit a low of $1.00 in 1985. Its price trajectory thereafter will be flat. In 1982, the national average retail price of No. 2 heating oil was $1.22 per gallon.

Large industrial users, however, can usually buy No. 2 oil at a smaller markup over crude compared to the typical home heating customer. In 1982, for example, the wholesale price of No. 2 oil available to large industrial customers was about $1.00 per gallon. That price will fall to a low of about $0.85 per gallon in real 1982 dollars and remain flat thereafter. NERA's projection of the wholesale price of distillate oil is shown—on an MMBtu basis—in Figure 11-13.

---

**American Metal Market*, May 23, 1983.

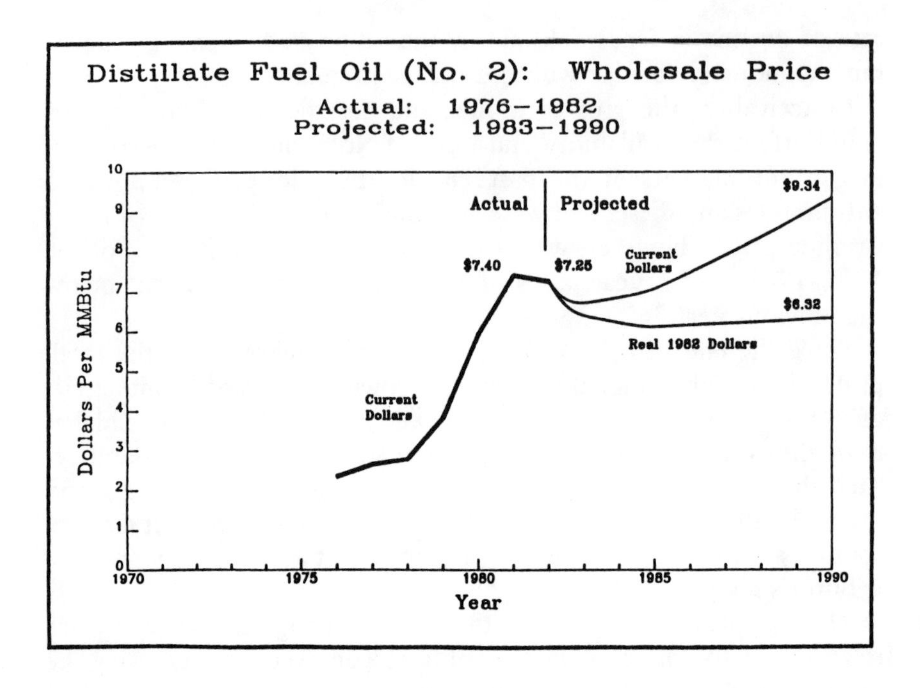


Figure 11-13

## Natural Gas

### *Supplies*

It appears that the future availability of natural gas supplies in the U.S. will be adequate for the rest of this decade and probably for the remainder of the 20th century. According to DOE's most recent official estimate, *proved reserves* in the lower 48 states are approximately 175 trillion cubic feet (Tcf). Producers explored for and proved up this gas at prices lower than those of today in real terms. There is, moreover, virtually unanimous agreement among industry analysts that the long-term future trend of gas prices is up. Therefore, this existing proved supply can be viewed as assured.

Beyond this amount is the category of *"inferred resources,"* the quantities of gas that exist in and adjacent to known fields. After proved reserves, this category is known with the greatest certainty. The U.S. Geological Survey estimates inferred resources in the lower 48 states to be 172 Tcf.* As these resources are associated with the proved reserves, the costs of their exploitation will not be significantly different from those of the proved reserves. Thus, there are an estimated 347 Tcf of gas known to exist with a high degree of certainty. These volumes would be economic to develop and produce at present prices.

Beyond these supplies are the large volumes of *undiscovered recoverable resources.* In the Survey's opinion, there is a 95 percent probability that the total undiscovered resources of natural gas in the lower 48 states are at least 422 Tcf. The margin of error in such estimates is unknown, but for illustrative purposes one can make the conservative assumption that the estimate overstates the true situation by 100 percent. This would reduce the figure for undiscovered resources to some 200 Tcf.

The Survey's estimated quantities of undiscovered resources are considered to be economic at the prices prevailing in 1980. Thus, even at today's much higher prices, not to mention future higher prices in today's dollars, this quantity of gas would be economic to find and produce.

In sum, there is more than 500 Tcf of resources in the lower 48 states from which future natural gas supply can be drawn, not counting such supplemental resources as tight sands and Devonian shales. This quantity is sufficiently great to permit the current level of production to continue into the 1990s without encountering resource constraints that would increase the real costs. Whether sufficient gas to satisfy a level of demand up to 20 Tcf per year will be found and produced is another matter. This will depend on two factors: the industry's perception of the resource position, and the expected future level of gas prices.

On balance, it appears that resources will allow gas production to continue at high levels over the next decade but not as high as the peak level of 22 Tcf per year experienced in the early 1970s. Domes-

*U.S. Geological Survey, *Estimates of Undiscovered Recoverable Conventional Resources of Oil and Gas in the United States* (Circular 860), 1981, Tables 1 and 6.

tic production will gradually decline from the 1981 level of about 19 Tcf to about 15 Tcf by 1990. This decline, however, will be partially offset by an increase in imports primarily from Mexico and Canada. Total supply including imports will be about 18 Tcf per year for the foreseeable future—a 10 percent decline from recent levels.

### *Gas Demand*

Through the 1980s, residential gas demand will grow modestly primarily because it will continue to be the preeminent fuel for home heating. During that period, the number of families and, thus, the number of households will grow—increasing the demand for gas. This, however, will be partially offset by the reduced consumption of gas per household due to conservation.

All of the other gas consuming sectors (i.e., commercial, industrial and electric utility) will show a decline in gas consumption through the 1980s. Although the economy will recover from the current recession, many of the basic industries that are major gas users (e.g., steel, automobiles, etc.) will not regain their past levels of output soon, if ever.

In addition, industrial customers will further conserve on their use of energy in general and gas in particular. Gas prices to industrial customers will rise relative to alternative fuels (see the discussion of gas prices below) causing many plants to switch to residual fuel oil. In addition, one of the biggest declines in gas consumption will occur in the electric utility sector because of the reduced rate of growth in the demand for electricity and an increased use of coal for generating electricity.

### *Gas Prices*

With respect to future gas prices, the greatest uncertainty is the politics of wellhead price deregulation. Will deregulation, as called for under the Natural Gas Policy Act (NGPA), be accelerated by Congressional action that moves up the presently scheduled end of price controls on new gas from the 1985 date? Will old gas be deregulated? Will there be any legislation abrogating provisions such as take-

or-pay and price escalation in current producer-pipeline contracts that now price new gas out of industrial end-use markets?

Given such uncertainties, the author has made the following three assumptions to estimate future gas prices: 1) controls on old gas will remain in force; 2) all new gas will sell at market-determined prices beginning in 1984; and 3) import deliverability from Canada and Mexico will be at 2.5 Tcf annually beginning in 1990, compared with the present availability of about 1.0 Tcf.

Using its energy modeling system, NERA projects (see Figure 11-14) that the national average wellhead price of natural gas in real terms will decrease from $2.79 per MMBtu in 1983 to $2.57 per MMBtu in 1985, rising thereafter to $3.53 in 1990 (i.e., in 1982 dollars). At first glance this seems inconsistent with the position stated above on gas supply—that the real price of gas will not be subject to upward pressure because of supply constraints.

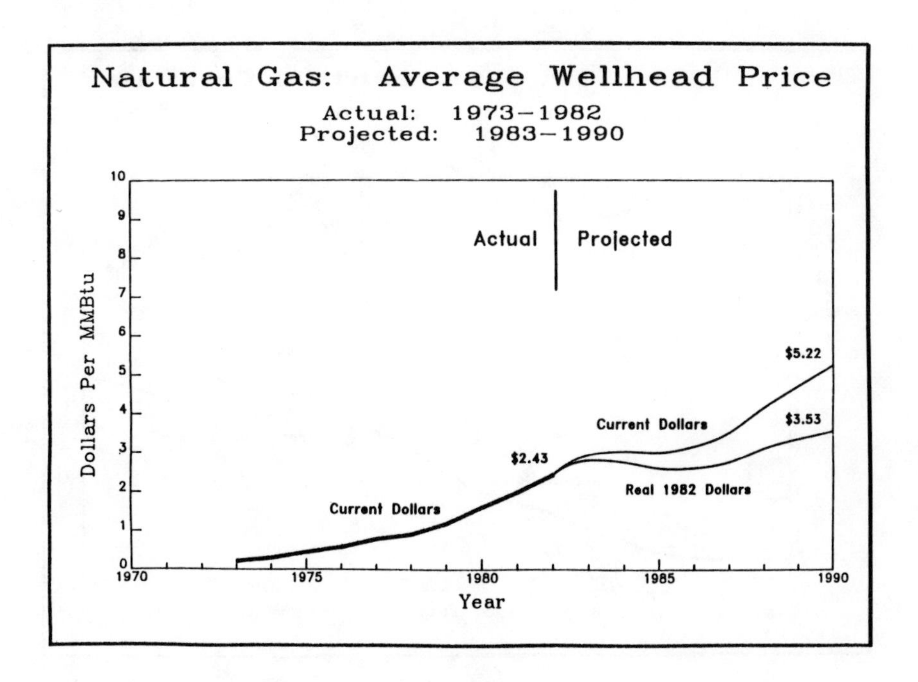


Figure 11-14

The price trajectory presented here, however, reflects several other changing circumstances. The price decline noted above constitutes the working off of the gas "bubble" (i.e., the immediate supply excess). The deregulated price of new gas after about 1985 will remain relatively constant. The "cheaper" old gas volumes, however, will be falling as a proportion of total available supply, so average price will be rising.

The above wellhead prices must be translated into burner tip prices by adding on transmission and distribution costs (see Figure 11-15). For industrial gas users, the burner tip price in 1985 will be $4.03 per MMBtu compared to $4.28 today. By 1990, the price to such users will increase to $5.05 per MMBtu (in 1982 dollars). Such price estimates, of course, will vary by location, particularly if gas utilities modify gas rate design at the pipeline or distribution company level in an effort to hold on to sales to industrial customers.

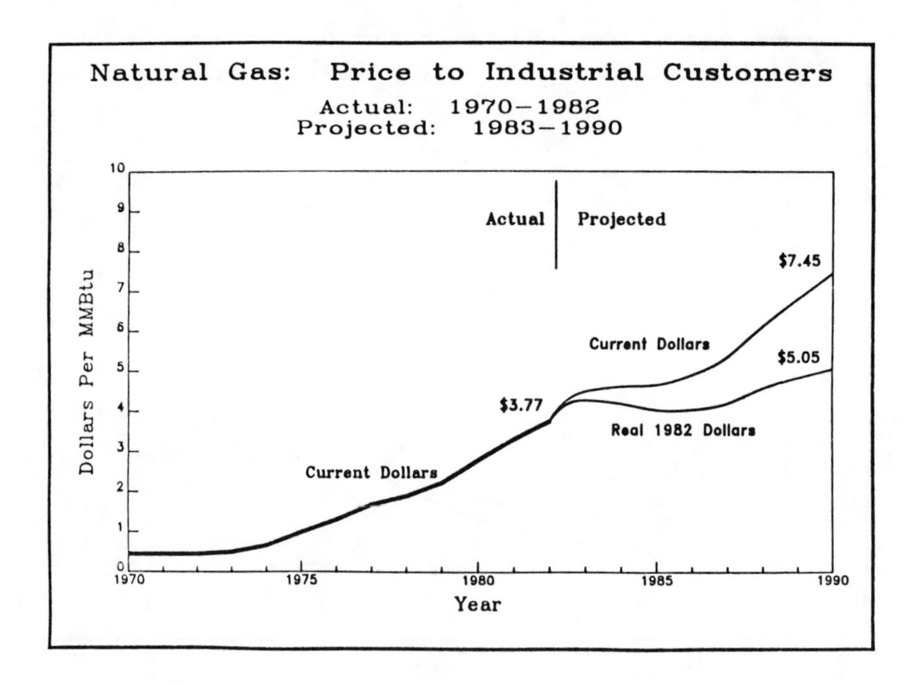


Figure 11-15

With the exception of natural gas, NERA's forecasts of fuel prices in real terms do not indicate any significant departure from current price levels. Fuel prices in current dollars, of course, will rise due to general inflation. This reflects a conviction that—again with the exception of natural gas—there are no economic forces that would move real fuel prices into new ranges, either above or below recent levels. Even allowing for healthy growth in demand, we see no great pressure on energy supplies—including natural gas—over the next decade.

Table 11-1 provides a comparison of prices of the various fuels discussed in this chapter on an MMBtu basis. These are national averages. Of course, oil products and natural gas costs vary widely across the country. Thus, the relative MMBtu costs of these fuels of a particular quality (e.g., sulfur content) and delivered to a specific cogeneration site could differ significantly from the costs shown in Table 11-1.

**Table 11-1. Prices of Selected Fuels to Industrial Customers**

| Fuel | Average Price 1981 | Average Price 1982 | Most Recent Price | NERA Forecast Price 1985 | NERA Forecast Price 1990 |
|---|---|---|---|---|---|
| | —(Dollars per MMBtu)— | | | (1982 Dollars per MMBtu) | |
| | (1) | (2) | (3) | (4) | (5) |
| No. 6 Residual Oil | $ 5.17[1] | $ 4.68[1] | $ 4.53[1] [2] | $3.95 | $4.09 |
| No. 2 Heating Oil | | | | | |
| Wholesale | 7.40 | 7.25 | 6.91[2] | 6.11 | 6.32 |
| Retail | 8.69 | 8.55 | 8.62[2] | 7.20 | 7.46 |
| Natural Gas | 3.32 | 3.77[3] | 3.94[2] [3] | 4.03 | 5.05 |

[1] Average retail price
[2] As of December 1982
[3] NERA estimate

Source: Columns (1)–(3): Lines 1, 2 and 3: U.S. Department of Energy, Energy Information Administration, *Monthly Energy Review*, May 1983.
Line 4: American Gas Association, *Gas Facts*, 1981.

## Conclusion

The price trends for the three cogeneration fuels are illustrated in Figure 11-16. The price of natural gas (as delivered to industrial users) rises above the price of resid in 1983, and must bear an increasing cost differential thereafter. In other words, "clean burning" gas has become more expensive on an MMBtu basis than resid (depending on sulfur content, of course). Thus, resid seems attractive as a cogeneration fuel because of soft world oil prices. Yet, even cheaper but "less clean" coal may undercut the fluid fuels in some cogeneration applications.

At the same time, the "softness" of energy prices is generally a boon to electric utilities. Moreover, they may be able to burn more and more coal at the margin, lowering those utilities' avoided costs. This could make cogeneration a less attractive proposition but the acid problem could scramble these comparisons.

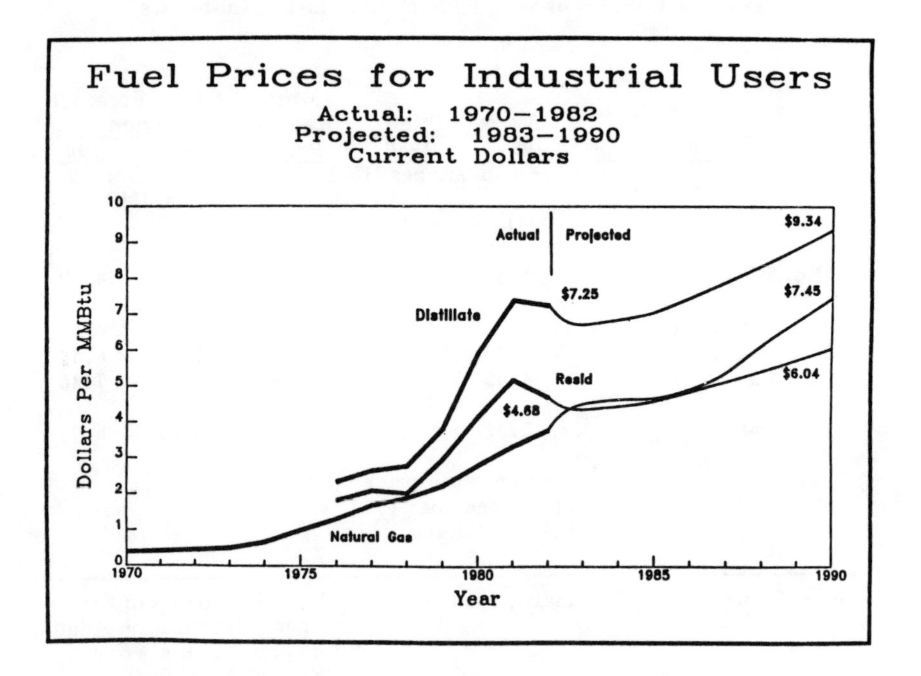


Figure 11-16

One thing is clear. Future fuel prices are tough to predict and uncertainty of all kinds abounds. Ignoring the risks inherent in such forecasts of future fuel prices and electric utilities' avoided costs is perilous for cogenerators. Nevertheless, such projections must be attempted because those numbers are crucial for cogeneration investment decisions. Thus, cogenerators must bet their money and take their chances.

# Chapter 12

# Environmental Regulations and Standards for Siting Gas- and Oil-Fired Cogeneration Systems

**_E. Wayne Hanson, Douglas D. Ober, and F. Richard Kurzynske_**

The following factors have together led to a renewed interest in natural gas- and oil-fueled cogeneration: 1) rapidly escalating energy prices, 2) the inherently high efficiency of cogeneration systems, and 3) recent favorable legislation at the federal and state levels. Currently, cogeneration is viewed by many as a significant business opportunity for the gas utility industry, as well as a cost- and energy-saving opportunity for the consumer. As a result, the market potential, especially for the smaller commercial and industrial applications (less than 10 mw), has been growing.[1] Environmental regulations and the permitting process for siting natural gas- and oil-fired cogeneration systems are not always readily understood. The purpose of this chapter is to present in concise terms a discussion of environmental concerns and an overview of environmental regulations and permit requirements for the siting of new gas- and oil-fired cogeneration systems.

Many of the potential industrial users of cogeneration systems (e.g., pulp and paper, chemical, and food processing) already have to comply with extensive environmental restrictions and regulatory requirements. Although the addition of a cogeneration system will not be subject to as many permitting requirements as a new major industry or utility, some cogeneration systems, especially in densely populated urban areas, may face increasingly strict environmental regulations.

The primary environmental regulatory concern is likely to be air quality. There may, however, be other regulatory requirements for noise, water quality, solid waste, and environmental impact assessments. In most cases, compliance with these other environmental regulatory requirements is easily achieved with inconsequential economic and operating impacts. In the case of air quality, significant reductions in emissions may be required in some areas of the country. These emission reduction requirements can usually be satisfied utilizing readily defined and available control technology.

In almost all situations, regulatory requirements exist that require cogeneration systems having emissions that exceed a stated threshold level to apply "Best Available Control Technology" (BACT) or state-of-the-art controls as determined on a case-by-case-basis. The threshold level that triggers BACT varies with the locale. Although additional emission standards are likely to be promulgated in the future, the application of BACT and requirements in some areas for emission "offsets" will be the dominant environmental regulatory requirements to be addressed in the siting of new gas- and oil-fired cogeneration systems. "Offsets" refer to a requirement that new sources attempting to locate in an area not meeting federal outdoor air quality standards must obtain reduced emissions from existing sources so as to show a net improvement in the outdoor air quality adjacent to the proposed site after the new source becomes operational.

## Environmental Concerns

The common environmental permitting concerns for internal combustion and stationary gas turbines relate to air pollution emissions and noise levels.

*Air Pollution Emissions*

Both reciprocating engines and gas turbines emit nitrogen oxides ($NO_X$), hydrocarbons (HC), particulates, sulfur dioxide ($SO_2$), and carbon monoxide (CO). Typical reciprocating engine emission factors, as published by the U.S. Environmental Protection Agency (U. S. EPA), are presented in Figure 12-1.[2] These emission factors are an aggregate of engine population statistics covering a broad range of duty cycles for engine sizes typically used in cogeneration applications. Two types of reciprocating engines are represented: gas-fired, spark-ignition (Otto cycle) engines and oil-fired, compression-ignition (Diesel cycle) engines.

Emissions and emission factors will vary considerably according to mode of operation and type of engine. Particulates and $SO_2$ emissions from diesel engines are primarily a function of proper engine operation and fuel quality, while they are negligible from gas-fired engines. CO emissions from diesel engines have been reported in the range of 0.3 to 14.6 grams/horsepower-hour (g/hp-hr).[3] Hydrocarbon emissions are less variable, but $NO_X$ emissions are reported to cover a wide range. For example, $NO_X$ emissions from gas-fired spark-ignition engines were reported by the U.S. EPA in 1979 to vary from 8 to 32 g/hp-hr. Recent reciprocating engine technology advancements promise $NO_X$ emission rates significantly below 8 g/hp-hr.

Figure 12-2 generalizes hourly emission rates (in pounds) for two engine sizes, again for Otto and Diesel types. As noted, the pollutant of primary concern is nitrogen oxides. The emission levels shown are intended only to provide perspective as they have been calculated by simply utilizing the emission factors shown in Figure 12-1 and do not represent a particular engine, account for any engine modifications that may have been made as a result of incorporating the engine into a cogeneration package, or reflect engine size differences. The U.S. EPA has reported reciprocating engines to be a significant contributor to the total $NO_X$ emissions from stationary sources (16.4 percent for 1971) and has projected stationary source $NO_X$ emissions in the U.S. to steadily increase through 1990.[3]

Emission factors developed by the U.S. EPA for gas-fired and oil-fired turbines are presented in Figure 12-3.[2] Figure 12-4 similarly

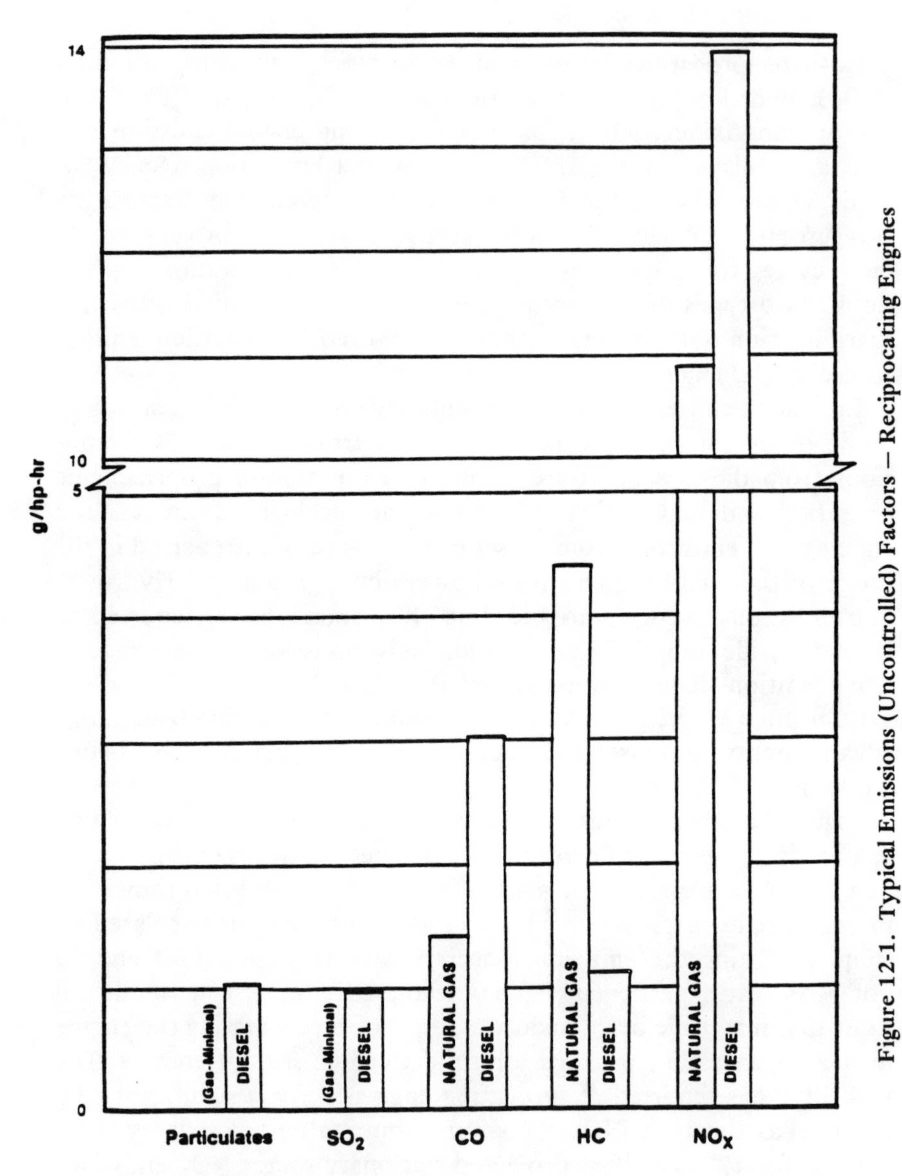


SOURCE: Compilation of Air Pollution Emission Factors AP-42 (Third Edition)

Figure 12-1. Typical Emissions (Uncontrolled) Factors – Reciprocating Engines

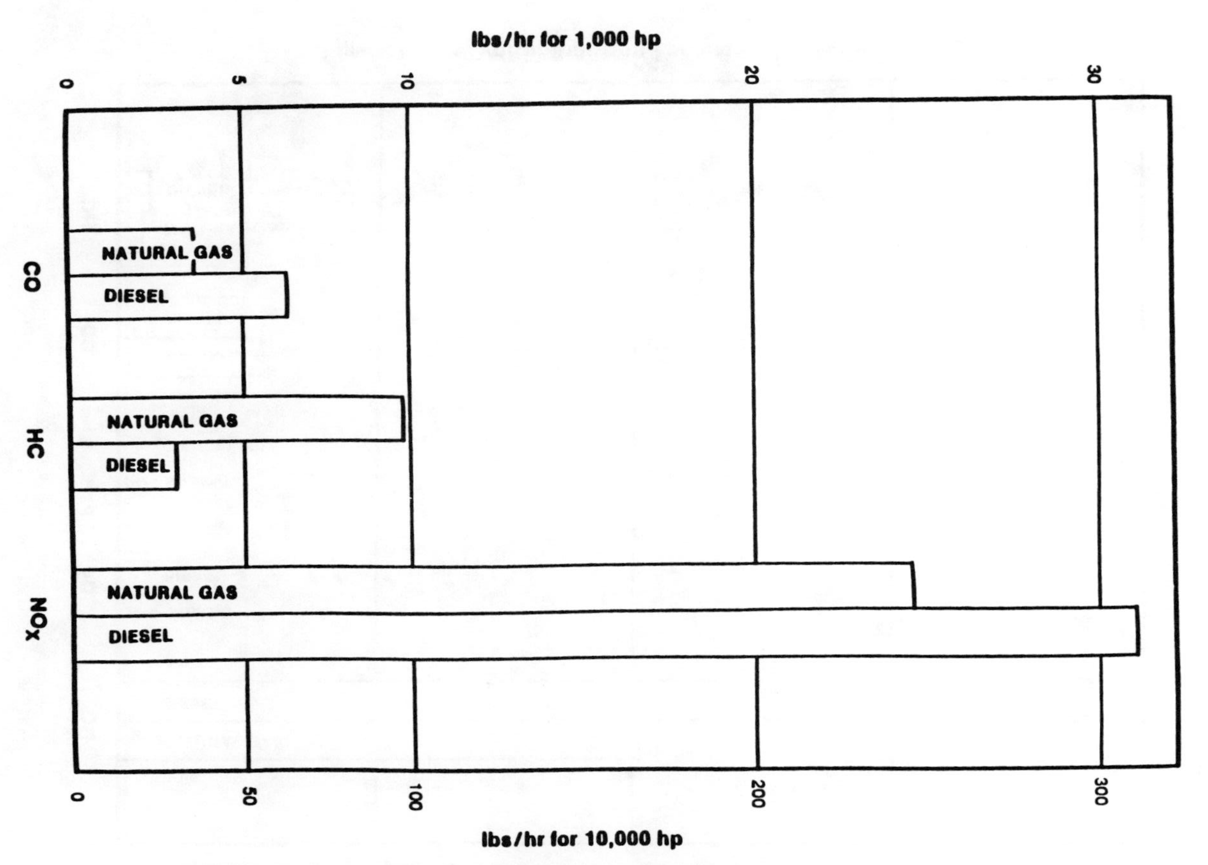


Figure 12-2. Reciprocating Engines Emission Rates (1,000 and 10,000 hp)

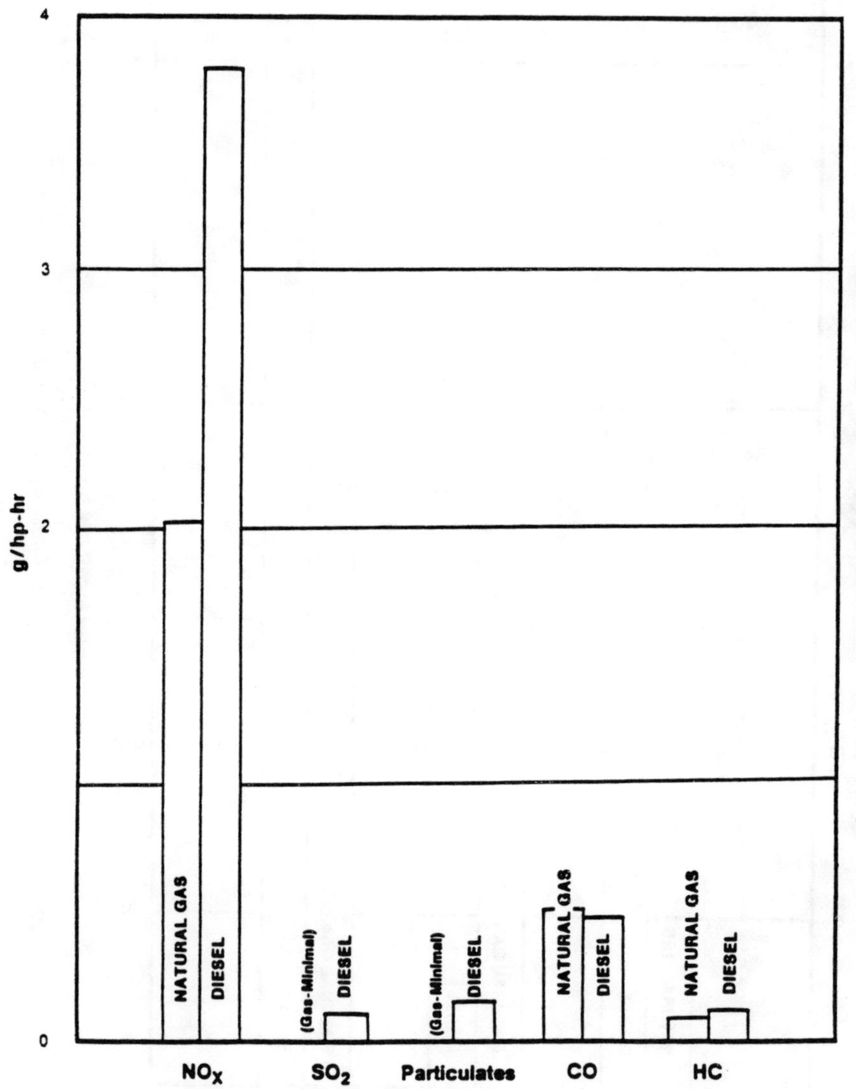


Figure 12-3. Typical Emission Factors — Gas Turbines

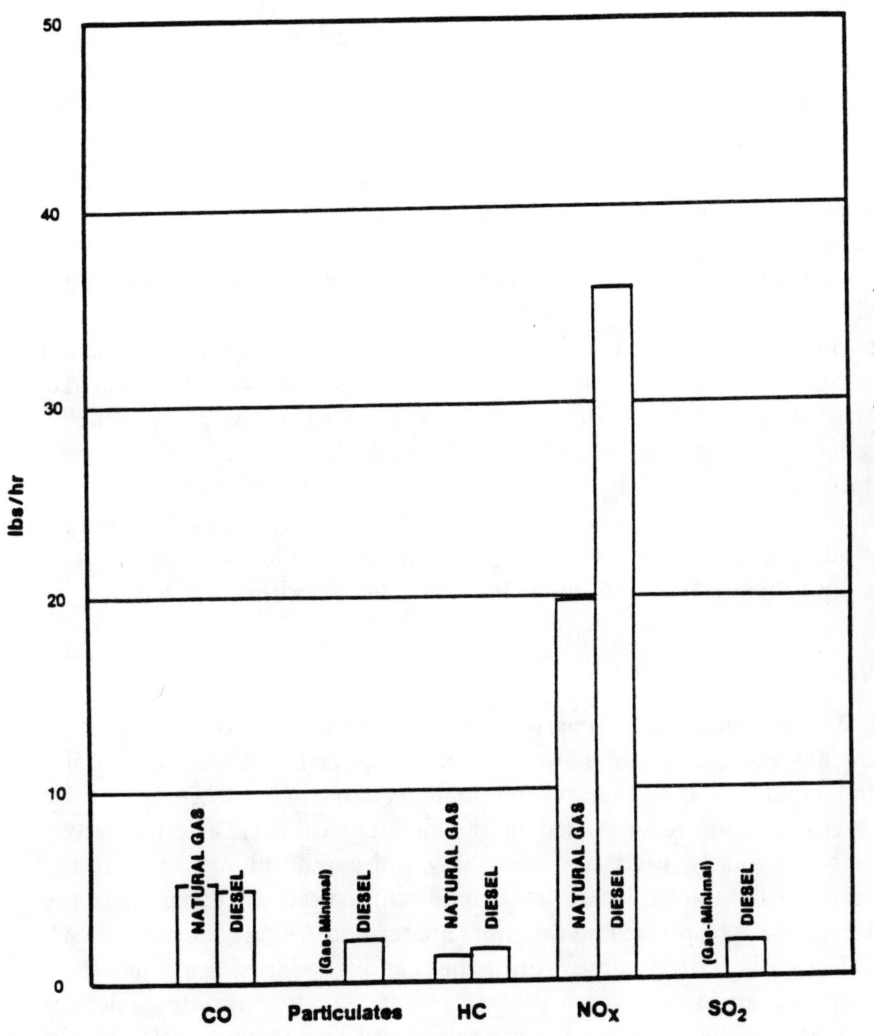


Figure 12-4. Gas Turbine Emissions (3 mw)

provides, based on the factors of Figure 12-3, some perspective of the emissions (pounds) for a typical 3 mw turbine generator. Since $NO_X$ emissions from gas turbines are less on a g/hp-hr basis than reciprocating engines and there are fewer stationary turbines in operation, the U.S. EPA recognizes that gas turbines contribute less (3 percent in 1974) to total stationary source $NO_X$ emissions than do reciprocating engines.[3] Emissions of $SO_2$ are determined by the sulfur content of the fuel and consequently are of little concern for natural gas-fueled units. Emissions of HC and CO are primarily determined by the efficiency of operation and have not been as significant an issue as $NO_X$.[4]

Cogeneration systems can benefit ambient air quality. For example, a cogeneration system with increased fuel efficiency (less fuel burned to meet electrical and thermal demands) is likely to have a lower emission rate than alternative systems. Likewise, the displacement of a coal- or oil-fueled energy system with a natural gas-fueled cogeneration system will typically decrease emission levels. Any such decrease realized may be considered when determining "offsets." Naturally, the overall air quality benefits will depend on the energy-producing technology replaced, the number and location of the newly located cogeneration units, and other site-specific considerations.

### *Noise*

Noise is an inherent part of the operation of both reciprocating engines and gas turbines. In addition to the prime mover noise, noise and vibration are generated by ancillary rotating machinery and by trucks or compressors used in the delivery of fuel. The regulatory unit commonly used for describing noise (sound pressure) is the decibel (dB). Noise levels are generally measured to approximate the human ear's response and the units are termed A-scale decibels (dBA). The A-scale is used in most ordinances and standards. For illustration purposes, an increase of 3 dBA or less is usually not detectable. An increase of 4 to 5 dBA is noticeable, and an increase of 10 dBA is perceived as a doubling of the loudness. Urban ambient noise typically ranges from 40 dBA to 80 dBA.[5] If the existing ambient noise is very low, a new source may be noticeable even if the absolute value of the noise levels meets the standard. In addition to dBA requirements, noise measurements and standards exist for smaller bands of sound frequency (octave bands) with center frequencies ranging from 31.5 hertz to 8,000 hertz. Uncontrolled cogeneration engine or turbine noise can range from 90 dBA to greater than 130 dBA when measured next to an operating unit. Effective noise attenuation has been developed for both reciprocating engines and gas turbines, utilizing inlet and exhaust silencers, as well as special enclosures or hoods, such that regulatory compliance is normally feasible.

### *Water Quality*

Water discharges from gas- and oil-fired internal combustion engines and turbine cogeneration systems are considered minimal. Reciprocating engines usually use a closed cooling system with no water discharge. The control of nitrogen oxides from gas turbines frequently involves the injection of specially treated water into the turbine combustor. The water pretreatment process can produce wastewater with a relatively higher concentration of dissolved solids. This effluent is usually acceptable to a public sewer system or, if this is not available, may be discharged to an evaporation pond or onsite treatment plant.

### *Solid Waste*

There is minimal solid waste associated with reciprocating engine and gas turbine cogeneration systems. Reciprocating engines require periodic oil changes, and the discarded oil may be sold to a reclaimer, leaving only disposal of discarded oil filter elements. Water effluent treatment or evaporation ponding may necessitate periodic disposal of small amounts of sludge.

## Environmental Regulations and Permit Requirements

### *Air Quality Emission Standards*

Two coupled approaches for protecting air quality have been established by the federal government. One approach is to place upon manufacturers of major pollutant-emitting equipment emission limits that the equipment they produce and sell must meet. The second approach is to place upon the equipment purchaser site emission limits that take into consideration the locale and the total emissions from all equipment installed.

Section III of the Clean Air Act authorized the U.S. EPA to promulgate regulations regarding standards of performance for new stationary sources. These regulations place emission limits on new major pollutant-emitting equipment and are referred to as New Source Performance Standards (NSPS). Such standards for stationary gas turbines were enacted September 10, 1979. As a result of petitions, the U.S. EPA proposed and adopted revised standards for stationary gas turbines on January 27, 1982. The revisions rescinded $NO_X$ emission limits for gas turbines greater than 30 mw baseload and placed $NO_X$ emission limits of 150 ppm on industrial and pipeline turbines less than 30 mw. The U.S. EPA's action effectively rescinded the requirement for water injection on industrial and pipeline turbines because of the unacceptable economic consequences of periodic shutdowns for needed internal inspections.(6) The new 150 ppm standard is to be achieved through dry control systems. However, the 75 ppm limit, which typically requires water injection, was retained for utility turbines of approximately 7.5 mw to 30 mw peak load. A simplified illustration of the revised NSPS for gas turbines is shown in Figure 12-5. Exemptions were allowed for natural gas-fired turbines when being fired with an emergency fuel and for regenerative (exhaust recuperated) cycle gas turbines less than 7.5 mw. All turbines are subject to an $SO_2$ emission limit of 150 ppm or a limit of 0.8 percent sulfur by weight of the fuel.

The 75 ppm and 150 ppm $NO_X$ limits can be proportionately increased for turbines of greater than 25 percent thermal efficiency (14.4 kiloJoules/watt-hour). In addition, when burning oil, the limit can be further increased to account for the nitrogen content of the fuels (0.015–0.25 percent, which adds 6–50 ppm $NO_X$ to the emission). All affected facilities must demonstrate compliance through field test results. The $NO_X$ emission standard and compliance demonstration test results are corrected to 15 percent $O_2$. The federal NSPS also contains monitoring requirements for fuel consumption, water-

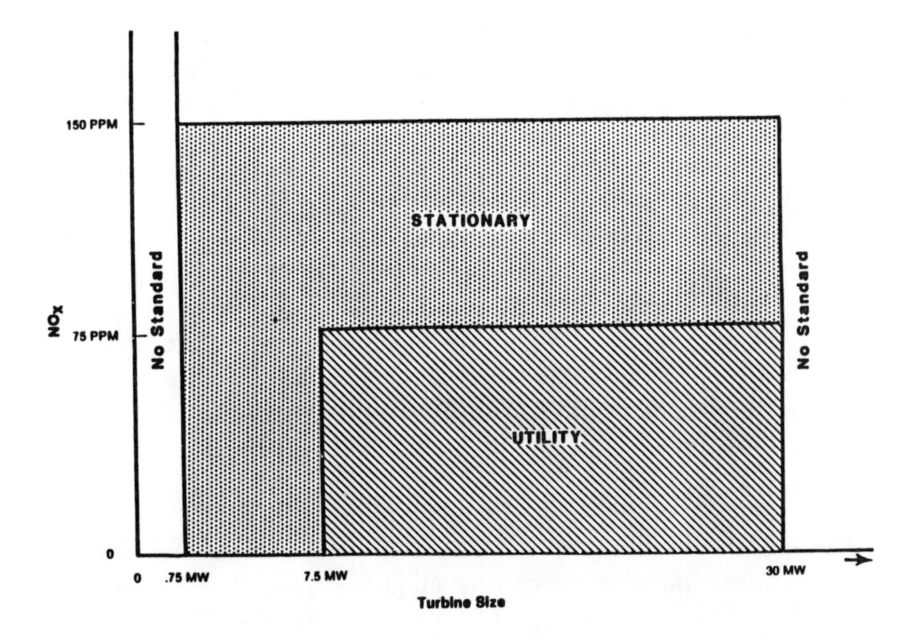


**Figure 12-5. EPA $NO_X$ NSPS for Stationary Gas Turbines**

to-fuel ratio (if using water injection), sulfur and nitrogen content of fuel, periods of water injection control deactivation due to ice fog, and use of emergency fuel (type, reasons, duration).[7]

The preamble of the Federal Register for the amended NSPS states that the turbine New Source Performance Standard is not intended to encourage or discourage cogeneration.[6] It mainly distinguishes between turbines with and without backup power. Cogeneration turbines can usually be supplied with backup power from the utility grid. Therefore, cogeneration units greater than 7.5 mw may be interpreted by some regulatory agencies to be subject to the 75 ppm $NO_X$ limit.

At the present time, New Source Performance Standards have not been promulgated for internal combustion reciprocating engines.

All state agencies are required to enforce regulations no less restrictive than the federally adopted standards. Therefore, all states have adopted the federal performance standards, usually by reference. In some cases, states have adopted emission standards for fuel-burning equipment, which by definition frequently includes gas turbines and internal combustion engines. Most of the fuel-burning equipment regulations address smoke, particulates, sulfur dioxide, and sulfur content of fuels, with generally a lesser emphasis on $NO_X$ emission limits.

Currently, the most restrictive local $NO_X$ emission standards appear to be in California. For example, San Diego County limits $NO_X$ emissions to 42 ppm at 15 percent $O_2$ for fuel-burning equipment fired on natural gas and 75 ppm for fuel-burning equipment fired on liquid or solid fuel.[4] These limits are applied to units with a maximum heat input rate of 50 million Btu/hour or more.

### *Air Quality Permit Requirements*

Depending on their size and location, new natural gas- and oil-fired cogeneration systems can be subject to federal, state, or local air pollution permit requirements. Permit requirements address site emission limits that take into consideration locale and total site emissions. Federal air quality permitting requirements may require a proposed cogeneration system to undergo a New Source Review (NSR) evaluation, which includes a Prevention of Significant Deterioration (PSD) review and permit application. NSR evaluations are triggered at the federal level by annual emissions of 250 tons/year of any criteria pollutant ($NO_X$, $SO_2$, particulate, CO, HC, $O_X$, and lead) for many sources, including diesel and gas turbines. As shown in Table 12-1, federally mandated NSRs do not typically apply to many of the smaller cogeneration systems.

As previously stated, state and local regulatory agencies must develop regulations and permitting requirements no less stringent than federally adopted standards. Environmental permitting requirements at the state and local level necessitate in all states the procurement of permits-to-construct through preconstruction reviews. The cutoff or threshold emission levels below which sources are exempt from preconstruction reviews vary. For example, some states require preconstruction permits for commercial fuel-burning equipment with a heat input rate of 1 million Btu/h or greater, whereas in other states

**Table 12-1. Maximum Size Cogenerator Not Requiring New Source Review**

| Technology | Megawatts | |
|---|---|---|
| Diesels | 30% efficient[a] | |
| $NO_x$ limit: | | |
| Oil-fired uncontrolled | 1.5 | |
| Dual-fired uncontrolled | 2.2 | |
| Proposed NSPS | 2.5 | |
| Gas turbines | 30% efficient | 20% efficient |
| $NO_x$ limit, assuming NSPS | 17.0 | 11.5 |
| $SO_x$ limit (oil-fired): | | |
| 1.0% sulfur oil | 5.0 | 3.3 |
| 0.3% sulfur oil | 16.5 | 10.8 |
| 0.2% sulfur oil | 24.8 | 16.3 |
| Steam turbines | 15% efficient | 10% efficient |
| $NO_x$ limit: | | |
| Coal-fired | 7.9 | 5.6 |
| Oil-fired | 3.4 | 2.4 |
| Gas-fired | 2.2 | 1.6 |
| $SO_x$ limit: | | |
| 0.2% sulfur oil | 5.0 | 3.3 |
| 1.0% sulfur oil | 1.0 | 0.7 |

[a] Plant electrical efficiency, BTU (electricity) 100/Btu (input fuel).
Source: Office of Technology Assessment.(1)

the cutoff level may be as high as 250 million Btu/h. In addition to preconstruction review requirements and permits, approximately 50 percent of the states require operating permits with or without annual fees. Such permits usually define emission limits and operating conditions as well as reporting and monitoring requirements.

Best Available Control Technology (BACT) and emission offset evaluations, when required, are performed during preconstruction reviews. The use of BACT is mandated at the federal level for those sources required to undergo NSR evaluations. State and local agencies have varying emission thresholds that trigger BACT regulations. For example, the California South Coast Air Quality Management District requires BACT if there is a net emission increase in $NO_X$ greater than 100 lb/day (effective January 1983).[8] In the San Francisco Bay Area, BACT applies if there is a net emission increase of 150 lb/day of $NO_X$.[9]

When triggered, BACT technology for an emission species is determined on a case-by-case basis and varies from area to area. BACT technology may be defined in terms of a type of emission control device required (for example, water injection for turbines), or BACT may be defined in terms of a maximum allowable emission rate. Presently, the San Francisco Bay Area has established BACT for gas-fired turbines as a maximum allowable $NO_X$ emission rate of 42 ppm at 15 percent $O_2$.

"Emission offset" requirements also exist in state and local permitting regulations. By law, a new source locating in an area having stringent emission regulations may be required to obtain a reduction in emissions from existing neighboring sources so as not to result in an increase in total emissions. The trigger level at which emission offsets are required varies from state to state and by local air pollution control agency. For example, within the State of California, which is considered to have the most stringent $NO_X$ emission regulations, emission offsets are required when the cumulative increase in emissions of $NO_X$ exceeds 550 lbs/day for the San Francisco Bay Area; 250 lbs/day for the San Joaquin Valley Air Basin; and 50 tons/year, 1000 lbs/day, or 100 lbs/hr, whichever is more restrictive, for the County of San Diego. Typically, cogenerators will go to extraordinary effort to avoid exceeding offset trigger thresholds. Such efforts include installing the most effective $NO_X$ reduction controls economically viable. Procurement of offsets can be costly and time-consuming.

### *Permitting Steps*

Since permitting regulation requirements can vary significantly between regulatory jurisdictions, a full delineation of permitting requirements can be made only on a site-specific basis. Early contact with the local permit-issuing agency is beneficial. If the cogeneration system to be installed has an emission level that triggers a permit requirement, early attention to acquiring the necessary permits can be crucial.

Typical periods required for state, local, and federal cogeneration air quality permit application preparation, review, and processing time are as follows:

*State/Local Permit*

| | |
|---|---|
| • Application preparation | 1 to 3 months |
| • Agency review, including additional information requests | 1 to 3 months |
| • Public notice/comments | 30 days |
| • Permit preparation | 15 to 30 days |
| Total estimated time | 3½ to 8 months |

The above estimated time assumes that limited ambient air quality analysis is required and offsets, if required, are readily available.

*Federal PSD Permit*

| | |
|---|---|
| • Application preparation (including screening modeling) | 1 to 3 months |
| • Monitoring (if required) | 12 to 16 months |
| • Refined modeling (including air quality related issues) | 1 to 3 months |
| • Agency processing time | 3 to 6 months |
| • Public notice/comments | 30 days |
| • Permit preparation | 30 days |
| Total estimated time, excluding monitoring | 7 to 14 months |

PSD permit processing time is generally in the 7-14 month time frame (excluding preconstruction ambient monitoring requirements). As the complexity of the location or detail of analysis increases, not only does the regulatory review time increase but also the time needed by the applicant to prepare the permit application. Although the PSD permit process is a federally developed program, a majority of states have assumed partial or complete responsibility for its administration.

### *Noise Standards*

There are no specific federal noise standards regulating industrial source impacts on the community. There are standards for the protection of workers [85 dBA average by the Occupational Health and Safety Administration (OSHA)] . Various other federal agencies, such as FHWA, EPA, HUD, and FAA, have published criteria noise levels used to evaluate particular projects. These criteria noise levels, however, will not typically impact most cogeneration projects.

Noise levels are generally regulated by local ordinances and, in some cases, by states. Regulatory requirements vary; for example, some regulate measured impacts at the nearest noise-sensitive receptor, such as a residence, school, church, hospital, convalescent home, etc. Others regulate noise impacts based on land-use zones of both the source and receptor, with impact measured at the property line. Such land-use related standards have been adopted by the State of Washington, the City of Portland, and San Joaquin County, California. Most standards are divided between day (7 a.m. to 10 p.m.) and night (10 p.m. to 7 a.m.) time periods. These standards are generally designed to prevent speech interference during the day and interference with sleep at night. Several of the regulations also have incorporated octave band standards, which are frequently applied to engine and turbine operations.

Figure 12-6 shows an example of octave band noise levels (uncontrolled) for a typical gas turbine.(10) Also plotted on Figure 12-6 is an example of Portland, Oregon's, daytime industrial noise standard.(11) As illustrated, noise attenuation may be necessary for gas turbine operations, particularly on low-frequency exhaust, and high-frequency casing noise at some locations.

### *Water Quality Standards*

Water discharges from gas- and oil-fired cogenerations are considered minimal and generally comply with existing water quality standards. Only on rare occasions must a cogenerator be concerned about water quality standards and procurement of water-related permits. Water quality permits are summarized in general terms as follows:

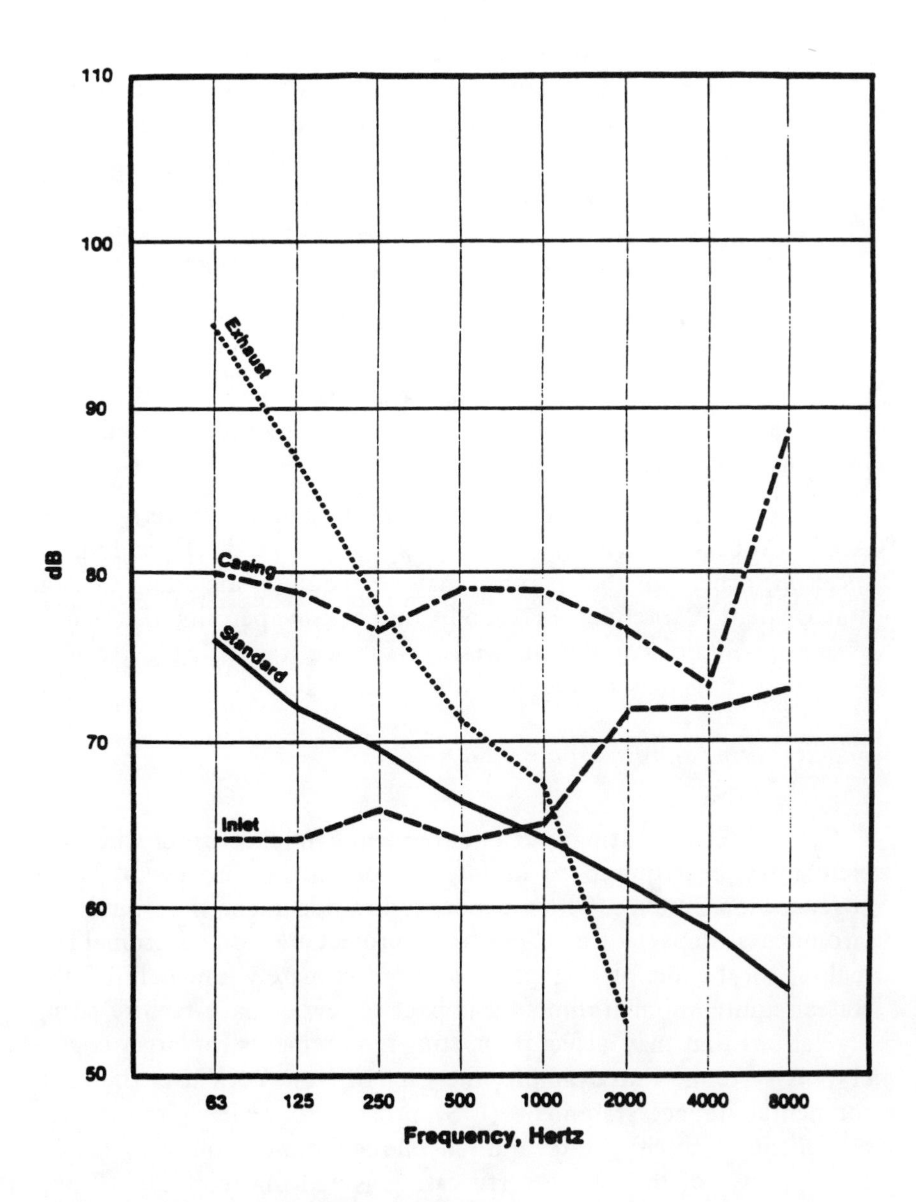


**Figure 12-6. Examples of an Industrial Octave Band Noise Standard and Levels for an Unattenuated Gas Turbine (Approximately 2,000 hp)**

- *NPDES Permit*—Regulates the discharge of pollutants from a point source into surface waters. Although this is a federal permit, it is frequently administered by the state.
- *State Waste Discharge Permit*—Regulates discharges into groundwaters or, in some states, into municipal sewer systems. If treatment prior to discharge is required, a waste treatment facility's approval or permit likely would be needed.
- *Reservoir Permit*—Usually regulates ponds or water storage over a defined size (e.g., depth of 10 feet).
- *City/County Permit*—Regulates a discharge permit and hookup charge to municipal sewer systems, frequently required by a city or county.

If a significant waste discharge is into navigable waters, a U.S. Army Corps of Engineers permit may be required. Depending on location, other possible permits may be required, including Coastal Management, Shorelines, or Flood Control Zone permits. If it is necessary to obtain water from wells or surface waters (e.g., streams), specific permits may be required on a state or local level.

### *Facility Siting and Environmental Impact Reviews*

Formal facility siting and environmental impact paper-study reviews may be required to initiate the permitting process of larger cogeneration systems. These paper studies document the overall environmental impact of a cogeneration project and add additional formality to the permitting process. Approximately one-half of the states require an environmental impact review or have facility siting regulations that may affect permitting requirements for larger cogeneration systems. For example, the City of New York uses the environmental impact statement (EIS) process to review projects for compliance with city, state, and federal regulations. Following review and approval of the EIS, a certificate is issued that contains all permit requirements. Most state environmental impact review or siting requirements pertain to larger cogeneration systems (i.e., ~50 mw); however, there are some states where these formal review requirements cover smaller facilities. Again, these requirements need to be evaluated on a site-specific basis.

*Proposed Air Quality Regulations*

A proposed federal regulation under development that could impact some cogeneration systems is the U.S. EPA's New Source Performance Standard (NSPS) for internal combustion (reciprocating) engines.[12] As originally proposed (July 23, 1979), the standard would limit $NO_X$ emissions and would apply to all new, modified, and reconstructed stationary internal combustion engines as follows:

- Diesel and dual-fueled engines greater than 560 cubic inch displacement per cylinder (CID/cyl).
- Gas engines greater than 350 CID/cyl or equal to or greater than eight cylinders and greater than 240 CID/cyl.
- Rotary engines greater than 1500 cubic inch displacement per rotor.

For diesel or dual-fuel engines, the limit as originally proposed is 600 ppm $NO_X$; for gas engines, the limit as originally proposed is 700 ppm. As with gas turbines, the $NO_X$ emission limit is expressed in ppm at 15 percent oxygen at standard atmospheric conditions. Again, a proportional increase is allowed for engines with thermal efficiencies greater than 35 percent (10.29 kJ/w-hr or 7,270 Btu/hp-h). The efficiency factor applies only to the engine itself and cannot be applied to the overall efficiency of a cogeneration system. No allowance has been given in the proposed standards for distillate fuels containing nitrogen. Monitoring requirements for diesel and dual-fuel engines may include the monitoring of: 1) intake manifold temperature, 2) intake manifold pressure, 3) rack position, 4) fuel injector timing, and 5) engine speed.

At this time, it is expected that the proposed NSPS will not initially pertain to natural gas-fueled engines, but only to diesel and dual-fuel engines. The timetable for promulgating this NSPS has been extended several times and is currently undetermined.[13]

*Future Trends*

Although future trends in regulatory requirements are not readily predictable, some speculation is possible. It is likely BACT requirements will continue to dominate or control most installation requirements. As bell weather agencies establish BACT requirements for cogeneration installations, other states and local agencies are likely to rely upon these initial BACT determinations and implement similar requirements.

## References

1. Office of Technology Assessment, Congress of the United States. *Industrial and Commercial Cogeneration.* OTA-E-192, February 1983.
2. U.S. EPA. *Compilation of Air Pollutant Emission Factors,* Third Edition, AP-42, August 1977.
3. U.S. EPA. *Stationary Internal Combustion Engines Standards Support and Environmental Impact Statement Volume 1: Proposed Standards of Performance.* EPA-450-2-78-125a, July 1979.
4. U.S. EPA. *Standards Support and Environmental Impact Statement Volume 1: Proposed Standards of Performance for Stationary Gas Turbines.* EPA-450/2-77-017a, September 1977.
5. Beranek, Leo (editor). *Noise and Vibration Control.* McGraw-Hill Book Co., 1971.
6. 47 *FR* 3767. January 27, 1982.
7. 40 CFR 60.33, *Subpart GG–Standards of Performance for Stationary Gas Turbines.*
8. Herb Lettice, South Coast Air Quality Management District. Personal communication.
9. Doug Wolf, Bay Area Air Quality Management District. Personal communication.
10. Ruston Gas Turbines, Inc. Personal communication to $CH_2M$-Hill, April 14, 1983.
11. City of Portland. Chapter 18–10, *Noise Control Ordinance.*
12. 44 *FR* 43152. July 23, 1979.
13. Doug Bell, U.S. EPA. Personal communication.

# Chapter 13

## Risk Management for Cogeneration Projects

***Thomas R. Germani***

One important part of the cogeneration planning process is the analysis of risks. Although many people believe that insurance is a subject which is separate and distinct from this risk analysis, they are wrong. Insurance is also thought of as one of the necessary evils in project development such as the obtaining of permits. This chapter shows that risk analysis and insurance are integral parts of the risk management process and can assist in project development.

### Risk Management

Before deciding to buy any insurance, the project's needs must of course be carefully analyzed. Risk Management, of which insurance is a part, is a process which can assist in the analysis and handling of the nonspeculative or pure risks presented by any endeavor. Nonspeculative or pure risks are defined as those risks facing the enterprise where there is only a chance of loss and no chance of financial gain. It is critical that a risk management review be done before deciding what insurances to purchase. Failure to follow this process could result in pure risk assumptions at worst, and, at best, major modifications in the insurance program as these risks become identified.

The four major steps of the Risk Management process are:

- Risk analysis
- Selecting the technique to handle each exposure

- Implementing the chosen technique
- Monitoring the decisions and implementing appropriate changes.

**Risk Analysis**

Risk analysis includes:

a. Identification of the potential risks of loss—property, liability, financial, etc.

b. Measurement of these risk exposures—frequency and severity.

**Selecting Techniques**

The selection of appropriate techniques includes:

a. Avoidance—Decide not to build the cogeneration plants.

b. Control—Exercise strict control over planning, design, construction, and operation of cogeneration facilities. (Quality Assurance)

c. Noninsurance Transfers

   (1) All elements of loss exposure—Have some other entity construct/operate the cogeneration facility.

   (2) Only potential financial impact—contractual—Shift liability for failures to contractors, suppliers, investors, engineer, etc.

d. Insurance—Purchase insurance for fortuitous events.

e. Retention—Retain risk and, depending on magnitude, establish a fund.

*The Next Step*

Just as a consulting engineer would be involved to assist with the technical aspects, or an investment banker for the financial aspects, insurance professionals should be employed to assist in the risk management aspects of the project.

**Selection and Role of Insurance Professionals**

The professional should be an insurance broker selected on the basis of:

- Experience with similar projects.
- Ability to understand the financial, legal and technical aspects of the project.
- Knowledge of the special insurances available.
- Capability to create a program to handle your project needs.

Once the insurance professionals have been selected, they should be fully integrated into the team. They should be involved in the conceptual planning, not just in financial studies, but, also, in plant design relative to combustion hazards, geographic and geophysical considerations, environmental problems and specific unique hazards. This will allow early identification of the risks and a determination as to which entity, project developer, investors, contractor, suppliers and/or purchaser of the output, should handle each of the risks. Second, the risk management team, including the financial advisors, legal advisors, engineering personnel and development staff, must work together to design a program so that the coverages will be adequate to take care of the risks involved in the project and so that the pricing will not destroy the project economics.

In many cases, risk management/insurance are not considered early in a project because of their perceived minimal impact. Risk Management is often misunderstood and insurance, in many situations, is a minor cost. In cogeneration facilities, as well as other project financed facilities, insurances have been developed which assume new and significant risks, and, therefore, are more expensive than traditional classes of insurance.

During the last three years, Johnson & Higgins has been involved in the risk identification process and the insurance negotiations for several cogeneration projects. Insurance programs have been arranged which range from the standard to the very unique. An early involvement in these projects was vital.

One question often asked is: When should the insurance broker be selected? One excellent answer is: four to six months before issuance of the contracts. This will allow adequate time to become familiar with the project and to assist in developing specifications for the applicable contract sections.

Once the insurance team has reviewed the technical, financial and legal aspects of the project, the structure of the organization and the experience of the participants who will design, construct and operate the facility, it is then appropriate to lay out the risks which need to be covered.

Depending upon the viewpoint, the risks/needs may be different. The purchaser of the output will want the project to be self-sufficient and does not want to obligate its credit. The lenders want to make certain that they receive their principal and interest, and the contractor wants to be sure that he is adequately compensated for his work. It is possible to provide all parties a certain degree of protection but there must be a sharing of risk by all participants. If there is no sharing, it is likely that the participants may lack sufficient incentive to do their utmost to make the project successful.

## Project Risks

All facilities which are project-financed present the same basic concerns. Will they be technically and financially sound? For a cogeneration project these concerns can be further defined:

- Can the facility be constructed *on schedule* and *within budget?*
- Once completed, will it *perform* as designed for the specified lifetime and within the *planned operating costs?*
- Will there be *sufficient input* to feed the plant?
- Can the output, steam and/or electricity, be sold at a *price* to help support the project?
- Can the project be *financed* so that the overall cost can be handled by its revenues?

Some of the factors that affect these risks are:

- Capability of the engineering/construction team.
- Adequacy of the system design.

- Ability of the developer to handle the diverse and intricate problems relative to siting, feedstock and energy sales.
- Federal, state and local regulations affecting construction and operation.
- Legislation/regulations affecting taxes and cost of energy, i.e. natural gas and oil supplies.
- General economic conditions.
- Ability of the developer to finance the project.
- Unforeseen events which may cause damage to the project, poor performance or delays.

The two listings above are the beginning of a risk analysis. Once a detailed analysis has been completed, using the input of all parties to the project, it is time to proceed to the second step of the Risk Management process—determine what techniques should be employed to treat each risk.

### Insurance Programs

The first technique, *avoidance,* probably would not apply. In order to avoid the risks of a cogeneration facility, one would have to decide not to build the facility. Since a cogeneration facility is intended to make more efficient use of energy resources, not building the plant is an unacceptable answer.

The four remaining steps will be utilized—*Control, Noninsurance Transfer, Insurance* and *Retention.* The ability to put together a viable insurance program depends upon the proper balance of these techniques.

Here is an overview of the insurances that would make up a typical insurance program. In general, there are three categories of insurance for cogeneration projects:

- *Standard*
  —Automobile Liability and Physical Damage
  —Boiler and Machinery
  —Comprehensive General Liability

—Property Damage (Fire and Extended Coverage, Builder's Risk and Permanent Plant Coverage)
—Umbrella Liability
—Workers' Compensation

These coverages are generally available at moderate cost.

- *Optional*
  —Business Interruption
  —Extra Expense
  —Performance Bonds
  —Property Damage ("All Risk," Difference in Conditions, Builder's Risk and Permanent Plant Coverage)
  —Transportation

Coverages such as these are more expensive, require tailoring to meet the needs of the project and must be carefully integrated with other coverages.

- *Special*
  —Contingent Business Interruption
  —Cost Overrun/Delayed Opening
  —Environmental Impairment Liability
  —Investment Tax Credit Recapture
  —Project Errors and Omissions
  —System Performance (Efficacy)

These insurances are relatively new, are available from a limited number of insurance companies and are more expensive than traditional coverages.

The first *two* groups of insurances can provide excellent coverage for losses resulting from damage to the facility and equipment destined for the project. These coverages can provide funds to the project to:

- Repair or replace equipment or property which is damaged. *(Property Damage)*
- Pay for continuing expenses while repairs or replacements are being made. *(Business Interruption)*

- Cover legal costs and judgments against the project for injuries to third parties. *(Comprehensive General Liability, Umbrella Liability)*
- Provide funds if a contractor/supplier fails to perform according to the contract. *(Performance Bonds)*

The standard and optional coverages will provide funds for the most likely exposures.

The third group of insurances provides coverage where failure of the facility to be completed on schedule, or to perform, is not due to physical damage, where there is physical damage to supplier's facilities or where the loss is due to professional errors, etc. These coverages can provide funds to the project to:

- Pay for continuing expenses if the project is delayed due to the inability of obtaining components because they or supplier's facilities have been damaged. *(Contingent Business Interruption)*
- Pay continuing expenses and/or increased capital costs caused by delays resulting from many "uncontrollable circumstances" which do not result in physical damage. *(Cost Overrun/Delayed Opening)*
- Make investors whole if the tax credits are recaptured by the Internal Revenue Service because the facility is destroyed. *(Investment Tax Credit Recapture)*
- Modify the plant so that it can meet specified performance levels. *(System Performance)*
- Pay continuing expenses while plant operates below specified performance levels. *(System Performance)*
- Cover consequential damages arising out of an error or omission of the project engineers. *(Project Errors and Omissions)*

## Insurance Marketing

The insurance industry is currently willing to offer much broader coverages than in the past for cogeneration facilities. Today the market is very soft and insurance companies are eager to increase their cash flow. Also, they are interested because the premium for the special coverages is quite high relative to the standard/optional coverages. This does not mean that all insurance companies are anxious to write these coverages. The underwriters' lack of understanding of the various technologies and the modification of traditional insurance conditions have caused the special insurances to be quite difficult to arrange.

The special insurances for cogeneration projects are expensive, take a long time to develop and require hard work by all parties. These insurances are not designed to, nor will they, make an uneconomic or technically unsound project feasible.

In order to be able to develop the comprehensive insurance program that cogeneration projects may require, one should understand the needs and concerns of the insurance markets, as well as those of the project. The following general rules should be kept in mind when looking for special insurances:

1. If the project is successful, everyone is rewarded, including the insurance company. If the project fails, everyone loses, or, at least, no one makes a profit. The exception to this is the bond purchaser or those providing loans. These receive their agreed upon return.

2. If the insurers, with a reasonable degree of accuracy, can predict that they will pay a loss during the policy period, the cost of the insurance will increase by the discounted value of the loss plus 30%.

3. Since the entire world capacity may be needed for each project, competitive quotations may not be possible.

4. Due to the amount of work involved in educating the insurance markets, it is prudent to wait until the major parameters of the project are resolved before formally approaching them.

5. The insurers are underwriting the capabilities of the participants as much as the technical and financial aspects of the project.

6. Extensive background information will be required so that the underwriters can be comfortable that the facility will be completed on schedule and will perform as projected.

7. In order to obtain the optimum insurance program, it is prudent to place all coverages with the same insurer or group of insurers, whenever possible. This approach will minimize gaps in coverage and potential difficulties in loss settlement.

Many of project needs can be met at a price that the project can afford. Two major areas where coverage is not being provided are:

- Guarantee of input.
- Price of the energy output.

However, if these, or other, risks for which there are currently no markets, can be adequately quantified, then it may be possible to develop insurance.

## Conclusion

At this time, Johnson & Higgins is working with several insurance companies which are very interested in insuring cogeneration projects. Given a viable project and a knowledgeable insurance team, the availability of insurance has not delayed any projects and, in some cases, has been the vital ingredient to allow them to move forward.

# Chapter 14

# Cogeneration/District Heating Implementation: A Developer's View

***Thomas R. Casten***

Implementing a district heating/cogeneration project is a long and complex process in which every step is interrelated to several other decisions. The process can be likened to solving a Rubik's Cube in which each turn affects several other faces and elements within each of these faces. This chapter offers some sense of the time involved and the key steps, and describes ways to solve the "district/cogeneration" cube to produce more profitable projects.

## Implementation—A Rubik's Cube

Each cogeneration project will have six major areas of decision or impact, just as a Rubik's Cube has six faces. The areas are:

- Technology Choice
- Thermal Loads
- Economic Feasibility
- Regulatory Approval
- Contracts
- Financing

Over the past eight years, and perhaps 400 project possibilities, CDC has learned that none of the choices are immediately obvious, and many first, second, and even third approaches are later replaced by more optimal technology, or by change of some other variable that turns a marginal project into an attractive project.

Each time an apparent snag is encountered, there is usually a way to solve it by changing several of the parameters in other areas, just as one changes the top face of a Rubik's Cube only by affecting several other faces. This may seem so obvious as to not bear mentioning, but in fact many would-be projects have been killed because the early feasibility assessments were performed with fixed assumptions about the technology choice, or the thermal loads, or the possible contract arrangements.

### Trenton District Energy Co.—A Case Study

The Trenton District Energy Company (TDEC) project history illustrates some of the interrelationships in producing a financeable, legal, economically attractive project. The Cogeneration Development Corporation (CDC) is the corporate general partner of TDEC which will own and operate the plant once it is built.

In 1977, Mayor Arthur Holland and his City Planning staff looked at their city and thought a downtown district heating plant powered by cogeneration might be feasible and might help revitalize their New Jersey State Capital City. They obtained Department of Energy grant money, put together a team to assess the project's merits, and spent a year producing quite an elaborate feasibility study.

The plan that emerged was to join roughly 20 State office buildings together into a *steam* district heating network powered by the cogenerated waste exhaust gas from four 2.5 mw combustion turbines which would generate 45 million kwh of electric power for sale to or use by the public utility, Public Service Electric and Gas Company (PSE&G). The turbines were to supply almost all of the heat, and were therefore to track the fluctuating thermal loads. This left the annual electrical load factor at only 51%.

It was concluded by some members of the development team that the project was not viable and they consequently dropped out of the team. The architect, Carl Stein, was reluctant to give up; he was convinced that a workable plan could be devised. The project in its initial technical conception employed a lot of capital for generating equipment, which did not operate very often because thermal loads were too low.

When CDC was asked to take over the development, it found a project that was apparently legal, was marginally economic, *and hence was not financeable.* However, there were a number of ways to further "twist the cube."

The first "twist of the cube" was to increase the system's thermal loads. The project engineers said steam could be transported economically for 2 miles, and a 350 bed hospital was found, located roughly 2 miles from the generating plant. It requires an extra million dollars to reach, and this seemed feasible. The new thermal loads would increase generation to over a 70% load factor, improve economics, and improve expected rate of return. But still the project faltered.

Technologically, steam was scrapped as a distribution medium and instead, high temperature hot water, using pre-insulated piping systems developed in Europe, was planned. The system had 100% return of cooled water, eliminated all steam traps, eliminated fear of rotting condensate pipes, and could be installed for only a slight increase in costs. The change to a hot water medium for distribution reduced projected distribution thermal losses *from 18% to 5%,* which was useful, but not the critical impact. With returning low temperature water, TDEC would no longer be restricted to the use of only high temperature exhaust heat recovery.

The project could now use reciprocating engines with their significantly higher electrical output and recover jacket water heat. Now the project was to have 12 mw electrical generation and could run at full electrical load, maximizing cogeneration. The projected load factor rose to 92%. The reciprocating engines have lower gross fuel rates per kwh but lower recoverable heat than combustion turbines. Possible recovery is 3100 Btu/kwh vs. 5800 Btu/kwh for the turbines.*

---

**Note:* The problem with reciprocating engines is that only 2100 Btu/kwh are recoverable from exhaust air steam send-out temperatures. The remaining 1000 to 2000 Btu per kwh come from the jacket water at 200°F, the oil cooler 160°F, and the after-cooler at 100°F. Only a small part of the thermal energy in the latter three sources can be used to preheat

This change would allow production of approximately 100 million kwh per year with full heat recovery.

Table 14-1 compares heat recovery with each technology to illustrate the dilemma.

**Table 14-1. Gross and Net Heat Recovery for the Turbine and IC Engines**
*All in Btu/kwh*

| | *IC Engine* | *Turbine* |
|---|---|---|
| Btu to Prime Mover per kwh Produced | 8,700 | 11,700 |
| Less Exhaust Heat Recovery (125 psig steam) | (2,120) | (4,280) |
| Net Fuel per kwh after Exhaust Heat Recovery | 6,850 | 7,420 |
| Net Fuel per kwh after Credit for 80% Efficient Boiler | 6,050 | 6,350 |
| *Other Heat Recovery Possible* | | |
| Jacket @ 200°F | 927 | 0 |
| Lube Oil @ 160°F | 458 | 0 |
| Charge Air @ 100°F | 600 | 0 |
| Possible Net Heat Rate | 4,595 | 7,420 |
| Possible Net Heat Rate after Credit for 80% Efficiency Boiler | 3,568 | 6,350 |

One more technological innovation was introduced that appeared to cause the "economics" face of the Trenton cube to twist into place, i.e. to become attractive and financeable. This last improvement of technology was to add supplementary firing to the exhaust heat recovery boilers. The engine exhaust would contain 11–12% free oxygen which could support further combustion at incremental fuel efficiencies of 95% (no need to heat ambient air to the boiler exhaust temperature, a normal fuel waster). The earlier plans had used conventional boilers operating at 82–85% efficiency for all heat above the engine recovery level. TDEC gained 10% fuel efficiency on 300 billion Btu of heat production.

---

feedwater in a stream system, whereas all jacket and some lube oil heat can be used in a hot water system.

**The Contract Face**

Developments at this stage moved to the "contracts" aspect of the cube solution, and suddenly TDEC was back to a near marginal project. Neither of the key thermal users, the State of New Jersey or the Mercer Medical Center could justify TDEC's requested starting thermal rates–rates which produced first year, second year, and even third year losses. It was now the "contract" face that had to be "twisted" for lowering the rate structure. Electrical sale was going nowhere, and PURPA rules were not yet out in New Jersey. TDEC was trying to sell directly to a municipal utility in Vineland, New Jersey, 60 miles away, and local politics were arcane. PSE&G regarded the project at this stage as highly unlikely and did not negotiate seriously.

New thermal loads saved the day and the project. An existing New Jersey State Prison was added as a customer. Although the prison was one mile beyond the far end of TDEC's then planned distribution, it had steady year-around loads. The prison had just committed to install 1500 tons of new high temperature absorption chillers that would use heat all summer. This addition allowed the economics to return to a financeable level, even with a lower thermal rate schedule. The State and Mercer Medical Center both agreed to sign 20-year contracts.

This step started to build political momentum and helped TDEC to reach agreement with Public Service Electric & Gas for a non-PURPA type sale of all electrical output under a rate formula which covered capital and O&M costs, plus tied fuel payments to 7800 Btu per kwh of TDEC's fuel, at TDEC's cost for fuel. The only tough spot, it seemed then–April of 1981–was that the contracts were "take as delivered" not "hell and high water." We had to somehow assure performance.

To assure performance, a panoply of contracts were necessary, and were finally executed. They included:

1. A stipulated sum turnkey construction contract backed by a 100% performance bond.
2. A 10-year maintenance contract from Cooper Industries, the engine vendor, which has a fixed ceiling on real maintenance

costs and which has fuel conversion guarantees—Cooper must pay for all fuel used by TDEC above a specified level of Btu per kwh.

3. Business interruption insurance.
4. A New Jersey State obligation to operate the project in the event of a major outage and to continue until the issues of bankruptcy were decided by a competent court.

With all these pieces in place on the "contract" face of the TDEC cube, financing could start to be firmed up. The United States Department of Housing and Urban Development (HUD) agreed to make a $4 million grant to Trenton for loan to TDEC at favorable rates. The New Jersey Economic Development Agency authorized the sale of $10 million of tax-exempt Industrial Revenue Bonds. The plan had been to raise another $3 to $5 million in private equity and complete the $17–$18 million financing.

### The "Cube" Has Other Faces That Move

Here in the TDEC story comes a clear indication of how the "cube" always seems to interrelate. In twisting the technical and thermal load face of the "cube" to add more users, more pipe and more generation, TDEC had pushed construction costs to $19 million. Debt service reserve, interest during construction, fees, etc., pushed total capital to $25 million. But projects that use tax-exempt revenue bonds cannot exceed $20 million expenditure over the six year period bracketing bond takedown. If TDEC could not keep the $10,000,000 bonds at tax-exempt interest rates, economics would slide again.

The solution finally adopted was a change in the financing structure to a lease of all generating equipment, essentially selling the tax benefits for equity. Fitting this lease to the by now complex project proved difficult enough, and was rendered much more difficult by narrow Industrial Revenue Bond (IRB) rules mandating that the lessor must be *"in the trade or business of leasing like or similar equipment."* Given that some aspects of TDEC are firsts in the USA, lawyers were greatly perplexed as to what lessor would qualify. Without clear qualifications, the leased equipment might be deemed a part of TDEC's capital expenditure and the bonds would become taxable. Many months were lost finding a lessor who would have no doubt about qualifying under IRB rules.

In August 1982, an outside force intervened and gave the financing face new life. The Tax Act was signed and it allowed district heating facilities to be exempt from the small issue test—allowed for unlimited use of the tax exempt Revenue Bonds to finance thermal distribution systems and user substations of district heating projects. Suddenly, the problem we had fought for months, spent hundreds of thousands of dollars solving, and had solved, became a nonproblem. Unfortunately, the new law only put in place one face.

### The Regulatory Face

In parallel, the regulatory "face" had to be solved and here, one particularly troublesome part of the regulatory "face," the Fuel Use Act (FUA), was found to be in place because of legislative oversight.

TDEC had absolutely no economic way to burn coal. New Jersey law requires 0.5% sulfur coal in Trenton which explains the complete absence of coal boilers in Trenton. The project, to be economic, had to use oil and/or gas, and seemed to be theoretically able to pass the economic exemption test of the FUA, but time was a killer. TDEC was told by the Energy Regulatory Agency that the FUA exemption process would take at least nine months.

Unknown to TDEC, when the switch from turbines to reciprocating engines was made, the project became exempt from FUA, as the FUA does not have jurisdiction over reciprocating engines. Once in a while, one gets lucky!

The state environmental permit required expensive dispersion modeling and 5 months to obtain—it was relatively easy. The Federal PSD permit was another story, taking a year which was characterized by the excessively slow administration of rules that were never designed for cogeneration. Just to name a few of the problems:

- No pollution credit was to be given for any of the boilers that would go out of service unless TDEC brought in certificates of destruction—1½ years before TDEC could offer service.

- All calculations had to be done with oil instead of gas because TDEC's gas supply is interruptible. This in spite of a history of only 8–10 hours annual interruption on the gas service contract in question.
- When all the dispersion modeling said that, in the very worst case, TDEC would meet standards with a 30-meter stack, the EPA ordered a 50-meter stack.
- When TDEC contemplated "twisting the cube" for different engines, we were told that any change in technology restarts the whole process.

**Final Dressing of the TDEC "Cube"**

Add a few random elements such as a rating from Standard & Poors (BBB+), a number of minor permits such as historical preservation, canal commission, building permits, and planning board permits, and the regulatory face seemed complete. Then it was discovered that because TDEC sells heat using "facilities granted by the State or a subdivision thereof," TDEC was subject to regulation by the New Jersey Board of Public Utilities.

Actually this had been known, the Board had been approached and in one of their many efforts to help TDEC, the Board declined to exercise their power to regulate. But an obscure part of the New Jersey Public Service rules said that no utility bonds of over 12 months life could be sold without approval by the New Jersey Board of Public Utilities. TDEC had to be approved. A friendly Board acted in three weeks, solving what was perhaps the last major regulatory hurdle.

There followed eight months needed to secure commitments on all financing. The process was slowed greatly by the new tax law as it opened up many opportunities for interpretation. By Christmas 1981, over five years since Mayor Arthur Holland had first begun his drive for district heating, it was finally clear that there could be a project—but only if the developer could obtain a private letter ruling on tax treatment of the leased property. Was TDEC a utility, or a non-utility? Would the fact that 100% of the present thermal contracts were with tax exempt institutions cause the IRS to deem TDEC a lease to the State of New Jersey?

The law firms representing the lender admitted that the issues raised had little likelihood of unfavorable ruling, but doubt existed. Lenders abhor doubt! With no comparable project started in the past 40 years, much of the relevant tax law had never been interpreted in tax court or by letter ruling. A crash effort was made and public figures' help was sought to obtain an expedited tax ruling. Senator Bill Bradley intervened, Congressman Ottinger wrote to the Treasury, HUD and DOE officials entered a plea, the United States and New Jersey Council of Mayors and the International District Heating Association added their weight.

**Lessons and Observations**

Before anyone concludes the TDEC experience is a normal timeframe for future projects, it must be pointed out that this has been from birth a cutting edge project, preceding many laws and their regulations that now ease the way. TDEC will not take five years to repeat, but it won't be repeated in five months either. Furthermore, the marriage of district heating with cogeneration raises some tax questions that would not arise for a pure cogeneration project, or district heating project.

There are several observations which can be offered to others developing cogeneration that are consistent with the TDEC experience, as well as our other cogeneration project experience.

**Lessons Offered**

1. Technical solutions improve with work. Don't stop with the first approach.
2. Try very hard to have the user(s) guarantee financing. Attempting pure project financing without such guarantees will easily add a year to the time. If you are one of the users, consider guaranteeing debt service.
3. Avoid adding time if possible. Law changes, even when they are beneficial in the long term, introduce "pendancy" and can stop a good project in its tracks.

4. Count on and cultivate political help. Cogeneration, especially with district heating, is quite attractive to most mayors.
5. Decide early what can be done with in-house staff, and when outside help must be sought. The most expensive course is to learn each lesson yourself.
6. It is extremely important to assemble a development team which has been through the process at least once and has had actual hands-on-experience. This means everyone from the engineers all the way to the bankers and lawyers.

There are many reasons to couple cogeneration with district heating and the rationale will grow stronger. Economics, energy conservation, lessened pollution and a vital electric utility system are all served by well-conceived projects. The road will get easier–has gotten easier–as more players enter the field. The economics are so profound that they ultimately place the noncogenerator at a competitive disadvantage. But, like everything else worthwhile, cogeneration takes a great deal of effort to develop.

They say there are some kids who can now solve the Rubik's Cube in less than a minute. The cogeneration/district heating "cube" will take longer, but the reward can be well worth the effort.

# About the Authors and Their Contributions

**Dilip R. Limaye** is President of Synergic Resources Corporation, Bala Cynwyd, PA. Chapter 1 is adapted from the author's presentation titled "Cogeneration in the U.S.: Problems and Prospects" at the American Power Conference, Chicago, IL, April 1982. His Chapter 3 material was presented at the Workshop on Planning Cogeneration Systems, co-sponsored by the International Cogeneration Society and Synergic Resources Corporation in New York, June 1983.

**James J. Zimmerman** is an independent consultant, and is the President of the Cogeneration Society of New York. His Chapter 2 was first presented at the Workshop on Planning Cogeneration Systems, co-sponsored by the International Cogeneration Society and Synergic Resources Corporation in New York, June 1983.

**Harry Davitian** is President of Entek Research, Inc., East Setauket, NY. Chapter 4 was adapted from a report prepared for the New York City Energy Office: "On-Site Generation in New York City," Entek Research, Inc., No. ERI 83/1, April 1983. The author alone is responsible for the material contained here.

**James R. Clements** is President of United Enertec, Inc. Material in Chapter 5 was presented at the Workshop on Planning Cogeneration Systems, co-sponsored by the International Cogeneration Society and Synergic Resources Corporation in New York, June 1983.

**Fred H. Kindl** is President of Encotech Inc., Schenectady, NY. Chapter 6 is adapted from a paper presented at the Workshop on Planning Cogeneration Systems, co-sponsored by the International Cogeneration Society and Synergic Resources Corporation in New York, June 1983.

**M.P. Polsky** is Supervising Mechanical Engineer and **R. J. Hollmeier** is Manager, Mechanical Engineering, at Fluor Engineers Inc., Power Division, Chicago, IL. Chapter 7 material was first presented as a paper at the Third International Conference on Cogeneration, co-sponsored by the International Cogeneration Society and the Electric Power Research Institute, Houston, TX, October 1983.

**J. M. Kovacik** is Manager, Cogeneration Systems Sales and Engineering, in the Industrial Sales Division, General Electric Company, Schenectady, NY. The material in Chapter 8 was first presented at the Texas A&M University Turbomachinery Symposium, Houston, TX, November 1983.

**M. V. Wohlschlegel** is Sales Manager, U.S. and Canada, **G. Myers** is Senior Engineer, Plant Development Engineering, and **A. Marcellino** is Senior Engineer, Plant Development Engineering, at Westinghouse Electric Corporation, Combustion Turbine Systems Division, Concordville, PA. The material in Chapter 9 was first presented at the Third International Conference on Cogeneration, co-sponsored by the International Cogeneration Society and the Electric Power Research Institute, Houston, TX, October 1983.

**Beno Sternlicht** is Board Chairman and Technical Director, Mechanical Technology Incorporated, Latham, NY. Chapter 10 was first presented at the Workshop on Planning Cogeneration Systems, co-sponsored by the International Cogeneration Society and Synergic Resources Corporation in New York, June 1983.

**Robert G. Uhler** is Vice President, National Economic Research Associates, White Plains, NY. Chapter 11 was presented at the Workshop on Planning Cogeneration Systems, co-sponsored by the International Cogeneration Society and Synergic Resources Corporation in New York, June 1983.

**E. Wayne Hanson** and **Douglas D. Ober** are with $CH_2$M-Hill, Inc., Portland, OR. **F. Richard Kurzynske** is Program Manager, Environmental Assessment and Control, Utilization, Gas Research Institute, Chicago, IL. Chapter 12 was presented at the Third International Conference on Cogeneration, co-sponsored by the International Cogeneration Society and the Electric Power Research Institute, Houston, TX, October 1983.

**Thomas R. Germani** is Vice President, Johnson and Higgins, New York, NY. Chapter 13 material was presented as a paper (© Johnson & Higgins 1983) at the Workshop on Planning Cogeneration Systems, co-sponsored by the International Cogeneration Society and Synergic Resources Corporation in New York, June 1983.

**Thomas R. Casten** is President, Cogeneration Development Corporation, New York, NY. Chapter 14 was presented at the Workshop on Planning Cogeneration Systems, co-sponsored by the International Cogeneration Society and Synergic Resources Corporation in New York, June 1983.

# Index

Note: Page numbers in italic indicate illustrations; page numbers followed by t refer to tabular material.